No-Frills Physics

No-Frills Physics

A Concise Study Guide for Algebra-Based Physics

Matthew D. McCluskey, PhD

CRC Press
Taylor & Francis Group
Boca Raton London New York

CRC Press is an imprint of the
Taylor & Francis Group, an **informa** business

CRC Press
Taylor & Francis Group
6000 Broken Sound Parkway NW, Suite 300
Boca Raton, FL 33487-2742

© 2019 by Taylor & Francis Group, LLC
CRC Press is an imprint of Taylor & Francis Group, an Informa business

No claim to original U.S. Government works

Printed on acid-free paper

International Standard Book Number-13: 978-0-367-21933-8 (Hardback)
978-1-138-58387-0 (Paperback)

Visit the Taylor & Francis Web site at
http://www.taylorandfrancis.com

and the CRC Press Web site at
http://www.crcpress.com

To my parents, John and Linda McCluskey

Contents

Preface

Physics is not easy. However, it *can* be simple.

This book is intended to help students learn basic physics principles, using a concise, problem-solving approach. It is a valuable resource for an introductory, algebra-based physics course. It is designed to supplement a traditional textbook and is not intended to be comprehensive or highly detailed. The point is to emphasize the most central topics and give the student practice in solving problems. Each chapter has three quizzes, which are closely connected to the examples in the text and problems at the end of the chapter. All answers are provided in the back of the book. Answers are usually given to two significant figures with the occasional rounding error. Note that instructors or online homework programs may require more precision.

For the sake of simplicity, I often omit units during the problem-solving steps. Be aware that some instructors want students to carry units through the derivation. I also commit the sin of not stating the units of certain fundamental constants (e.g., permittivity). However, the list of physical constants in the back of the book contains units in all their glory.

I would like to thank Anya Rasmussen for extensive comments on the text and Everett Lipman for valuable discussions. I am indebted to Jesse Huso, Slade Jokela, Chris Pansegrau, Violet Poole, Anya Rasmussen, and Jacob Ritter for proofreading and senior editor Lou Chosen for initiating and guiding this project.

Corrections or comments are welcomed and may be sent by e-mail to: mattmcc@alum.mit.edu

Author

Matthew D. McCluskey is a professor in the Department of Physics and Astronomy and Materials Science Program at Washington State University (WSU), Pullman. He received a Physics PhD from the University of California, Berkeley, in 1997, and was a postdoctoral researcher at the Xerox Palo Alto Research Center (PARC) in California from 1997 to 1998. Dr. McCluskey joined WSU as an assistant professor in 1998. His research interests include semiconductors, high-pressure physics, and optics. Professor McCluskey has taught WSU's introductory algebra-based physics courses numerous times.

Motion

POSITION AND VELOCITY

To describe the position of an object, we use an axis (Figure 1.1).

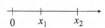

FIGURE 1.1 A one-dimensional axis.

The 0 point is the *origin*. Assume an object is at a position x_1, measured in meters from the origin. A while later, it is at position x_2. The *displacement* (Δx) is the distance between these two points:

$$\Delta x = x_2 - x_1$$

The time it takes to go from x_1 to x_2 is the *time interval*, Δt. Assume the object's motion is steady (not slowing down or speeding up). The *velocity* (v) is measured in meters per second (m/s):

$$v = \Delta x / \Delta t$$

1. A car is at a position $x_1 = 10$ m and travels at a constant velocity; 3.0 s later, it is at $x_2 = 40$ m. What is the displacement and velocity?

   Displacement $= \Delta x = 40$ m $- 10$ m $= \mathbf{30\ m}$

   Velocity $= v = 30$ m/3 s $= \mathbf{10\ m/s}$

2. A plane is at a position $x_1 = 100$ km and travels at a constant velocity; 500 s later, it is at $x_2 = 50$ km. What is the displacement and velocity?

Displacement $= \Delta x = 50$ km $- 100$ km $= \mathbf{-50\ km}$

Velocity $= v = -50$ km/500 s $= \mathbf{-0.1\ km/s\ or\ -100\ m/s}$

ACCELERATION

Acceleration (*a*) is a change in velocity (Δv) over a change in time (Δt):

$$a = \Delta v / \Delta t$$

It is measured in units of meters per second, per second, or m/s^2.

If a is positive, then velocity is increasing. This is what happens to a car when you step on the accelerator pedal. If acceleration is negative, then your velocity is decreasing, like when you slam on the brakes. If $a = 0$, then Δv is 0, which means that the velocity is constant.

Figure 1.2 shows a plot of a car's velocity, where t is the time. Initially, v is increasing, so acceleration is positive ($a > 0$). Later, v is constant, so $a = 0$. Finally, the car slows down, so acceleration is negative ($a < 0$).

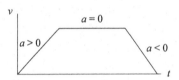

FIGURE 1.2   Plot of velocity versus time.

1. A car starts from rest. After 5.0 s, it has a velocity 20 m/s. Find the acceleration.

$\Delta v = 20$ m/s

$\Delta t = 5$ s

$a = 20$ m/s/5 s $= \mathbf{4\ m/s^2}$

2. A plane travels with a velocity 40 m/s. After 10 s, it has a velocity 100 m/s. Find the acceleration.

$\Delta v = 100$ m/s $- 40$ m/s $= 60$ m/s

$\Delta t = 10$ s

$a = (60 \text{ m/s})/(10 \text{ s}) = \mathbf{6.0 \ m/s^2}$

QUIZ 1.1

1. A runner is at a position $x_1 = 25$ m and has a constant velocity. After 9.0 s, she is at a position $x_2 = 70$ m. What is the displacement and velocity?
2. A ball is at a position $x_1 = -10$ cm and has a constant velocity. After 5 s, it is at a position $x_2 = -50$ cm. What is the displacement and velocity?
3. A missile starts from rest. After 10 s, it has a velocity 500 m/s. Find the acceleration.
4. A car travels with a velocity 20 m/s. The driver presses on the brakes, and 5.0 s later, the car has stopped. Find the acceleration.

VELOCITY AS A FUNCTION OF TIME

We use equations to predict an object's position and velocity at some later time. Suppose we have a stopwatch. We start at time $t = 0$. The position at $t = 0$ is the *initial position* (x_0). The velocity at $t = 0$ is the *initial velocity* (v_0). At a later time, t, the velocity is given by

$$v = v_0 + at$$

Note that v is the *instantaneous velocity*. It is given by $\Delta x / \Delta t$, where Δt is small such that the velocity does not change significantly during the time interval.

1. The initial velocity of an object is 5.0 m/s, and the acceleration is 3.0 m/s². What is the velocity at $t = 7.0$ s?

$v_0 = 5$ m/s

$a = 3$ m/s²

$v = 5 + (3)(7) = \mathbf{26 \ m/s}$

2. An object starts from rest and accelerates at −7.0 m/s². Find the velocity after 11 s.

$v_0 = 0$

$$a = -7 \text{ m/s}^2$$

$$v = 0 + (-7)(11) = -77 \text{ m/s}$$

POSITION AS A FUNCTION OF TIME

The object's position (Figure 1.3) is given by:

$$x = x_0 + v_0 t + \tfrac{1}{2}at^2$$

FIGURE 1.3   An object's position is given by x.

1. An object has an initial position 2.0 m, initial velocity 3.0 m/s, and acceleration 4.0 m/s². What is the position at $t = 5.0$ s?

$$x_0 = 2 \text{ m}$$

$$v_0 = 3 \text{ m/s}$$

$$a = 4 \text{ m/s}^2$$

$$x = 2 + (3)(5) + \tfrac{1}{2}(4)(5^2) = 67 \text{ m}$$

2. An object starts from rest and accelerates at 5.0 m/s². What is the displacement after 10 s?

   Let's have the object start at the origin. This simplifies the problem. Because the object starts at the origin, $x_0 = 0$.
   The object starts from rest, so $v_0 = 0$.

$$a = 5 \text{ m/s}^2$$

$$x = 0 + (0)(10) + \tfrac{1}{2}(5)(10^2) = 250 \text{ m}$$

QUIZ 1.2

1. The initial velocity of an object is −5.0 m/s, and the acceleration is 1.0 m/s². What is the velocity at $t = 5.0$ s?

2. An object starts from rest and accelerates at 10 m/s². What is the velocity after 3.0 s?

3. An object has an initial position −6.0 m, initial velocity −5.0 m/s, and acceleration 4.0 m/s². What is the position at $t = 5.0$ s?

4. An object starts from rest and accelerates at −0.10 m/s². What is the displacement after 10 s?

TWO OBJECTS

Sometimes we want to track the position of two objects (Figure 1.4). We will write the equation of motion as

$$x = x_0 + v_0 t + \tfrac{1}{2}at^2$$

for each one individually. To find when the two objects are at the same position, we set their x values to be equal.

FIGURE 1.4 Two objects have two different x values.

1. Kevin walks with a velocity 1.0 m/s. He is initially 300 m ahead of Rodney, who runs at a velocity 6.0 m/s. How long does it take for Rodney to catch up with Kevin?

Assume constant velocity ($a = 0$).

Kevin: $x = 300 + (1)t + \tfrac{1}{2}0t^2 = 300 + t$

Rodney: $x = 0 + (6)t + \tfrac{1}{2}0t^2 = 6t$

Set them equal: $300 + t = 6t$

$300 = 5t$

60 s $= t$

2. A missile is launched at a jet, which is initially 250 m away. The missile accelerates from rest, with $a = 100$ m/s². The jet has a constant velocity 200 m/s, away from the missile. How long does it take for the missile to impact the jet?

Missile: $x = 0 + 0t + \tfrac{1}{2}(100)t^2$

Jet: $\qquad\qquad x = 250 + (200)t + \tfrac{1}{2}\,0t^2$

Set them equal: $50t^2 = 250 + 200t$

$$50t^2 - 200t - 250 = 0$$

Quadratic equation: $t = \dfrac{200 \pm \sqrt{40{,}000 + 4(50)(250)}}{2(50)} = \dfrac{200 \pm 300}{100} = -1.5$

Only the positive answer makes sense, so $t = \mathbf{5.0\ s}$

GRAVITY

Gravity causes all objects to accelerate downward. This *gravitational acceleration* is denoted as g. On Earth, $g = 9.8$ m/s². (We will always ignore air resistance.)

Let's choose an axis that points up. Then gravitational acceleration has a *negative* value, -9.8 m/s². That's because gravity points down, which is opposite to our axis.

1. An object is dropped from a height 20 m (Figure 1.5). How long does it take to hit the ground?

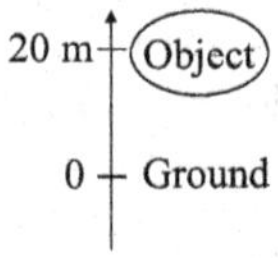

$$20\ \text{m} + \boxed{\text{Object}}$$
$$0 + \text{Ground}$$

FIGURE 1.5   An object is dropped from a height of 20 m.

$x = x_0 + v_0 t + \tfrac{1}{2}at^2$

$x_0 = 20$ m

$v_0 = 0$ ("dropped" or "released" means the initial velocity is 0)

$a = -9.8$ m/s²

$x = 0$ when the object hits the ground. Find t.

$0 = 20 + 0t + \tfrac{1}{2}(-9.8)t^2$

$$0 = 20 - 4.9t^2$$

$$4.9t^2 = 20$$

$$t^2 = 4.08, t = \textbf{2.0 s}$$

2. A person throws a ball upward, with an initial velocity 15 m/s. The ball goes up and comes back down. How long is the ball in the air before the person catches it?

Let $x_0 = 0$

$x = 0$ when the person catches the ball.

$$0 = 0 + 15t + \tfrac{1}{2}(-9.8)t^2$$

$$0 = 15t - 4.9t^2$$

$$0 = 15 - 4.9t$$

$$4.9t = 15, t = \textbf{3.1 s}$$

QUIZ 1.3

1. A mouse runs away from a cat at a velocity 2.0 m/s. The cat is initially 10 m away from the mouse. The cat runs toward the mouse at 7.0 m/s. How long does it take for the cat to catch the mouse?
2. A bank robber drives a car at a constant velocity 20 m/s. A super-fast police car is initially at rest 800 m away. The police car accelerates at 20 m/s². How long will it take for the police to catch the robbers?
3. A ball is dropped from the roof of a building 44 m above the ground. How long does it take for the ball to hit the ground?
4. A person shoots a paintball straight up. The initial velocity is 99 m/s. How long does it take for the paintball to hit the ground? (Ignore the height of the person.)

CHAPTER SUMMARY

Displacement (m)	$\Delta x = x_2 - x_1$
Velocity (m/s)	$v = \Delta x / \Delta t$
Acceleration (m/s²)	$a = \Delta v / \Delta t$
Velocity as a function of time	$v = v_0 + at$
Position as a function of time	$x = x_0 + v_0 t + \tfrac{1}{2}at^2$
Gravitational acceleration	$g = 9.8$ m/s², down

END-OF-CHAPTER QUESTIONS

1. An object has an initial velocity of 8.0 m/s. It slides to a stop in 2.0 s.
   a. Find the acceleration.
   b. Find the displacement.
2. An object starts from rest and accelerates at 10 m/s². What is the displacement after 10 s?
3. In a sprint practice, Boris runs at a constant velocity 8.0 m/s. Fred is initially 20 m behind Boris. Fred runs at a constant velocity 10 m/s. How long will it take for Fred to catch up with Boris?
4. A water balloon is dropped from the roof of a building 123 m above the ground. How long does it take for the balloon to hit the ground?
5. An object is launched upward from the ground with an initial velocity of 35 m/s. How long is the object in the air?

ADDITIONAL PROBLEMS

1. An object accelerates from rest, $a = 4.0$ m/s².
   a. How long does it take to travel 2.0 m?
   b. What is its velocity, after it has traveled 2.0 m?
2. Two cars approach each other. Initially, they are 375 m apart. The first car has an initial velocity of 10 m/s and accelerates at 20 m/s². The second car has constant velocity of –15 m/s.
   a. How long will it take for them to collide?
   b. What is the displacement of the first car when the collision occurs?
3. A ball is released from a height of 8.3 m.
   a. How long does it take for the ball to hit the ground?
   b. What is the velocity of the ball just before it hits the ground?
4. A ball is thrown upward with an initial velocity 7.1 m/s. How long does it take the ball to reach its maximum height?

Vectors

CARTESIAN COORDINATES

In Chapter 1, we looked at motion along one dimension, x. Now we are going to look at two dimensions, x and y. We plot an object's position (x, y) using Cartesian axes (Figure 2.1).

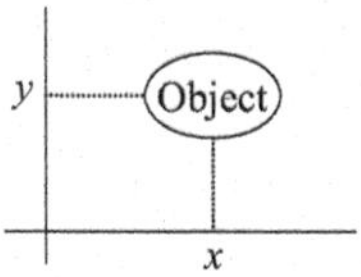

FIGURE 2.1   An object's position in a plane described by x and y.

We need two numbers, x and y, to describe the position. Similarly, we need two numbers, v_x and v_y, to describe the velocity. Consider a constant velocity. An object's x value changes by Δx in a time Δt, while the y value changes by Δy.

The x component of the velocity is

$$v_x = \Delta x / \Delta t$$

The y component of the velocity is

$$v_y = \Delta y / \Delta t$$

1. A car is at a position (10 m, 15 m) and travels at a constant velocity; 3.0 s later, it is at (16 m, 3.0 m). What is the velocity?

$$\Delta x = 16 \text{ m} - 10 \text{ m} = 6 \text{ m}$$

$\Delta y = 3 \text{ m} - 15 \text{ m} = -12 \text{ m}$

$v_x = 6 \text{ m}/3 \text{ s} = \textbf{2.0 m/s}$

$v_y = -12 \text{ m}/3 \text{ s} = \textbf{-4.0 m/s}$

2. An object travels at a steady speed. It moves 6.0 m toward the left in 0.025 s. What was its velocity?

$v_x = -6/0.025 = \textbf{-240 m/s}$

$v_y = 0/0.025 = 0$

VECTOR COMPONENTS

Velocity is an example of a *vector*. One can visualize a vector by drawing a right triangle (Figure 2.2). The horizontal leg is v_x, and the vertical leg is v_y.

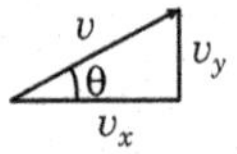

FIGURE 2.2 A velocity vector has x and y components, v_x and v_y.

The vector, an arrow that points in a specific direction, is the hypotenuse. The length of the vector is its *magnitude* and is given by

$$v = \sqrt{v_x^2 + v_y^2}$$

The magnitude of a velocity vector is called *speed*.

A vector has a magnitude and direction. We can express direction by an angle θ with respect to horizontal. From trigonometry,

$$v_x = v \cos\theta \quad v_y = v \sin\theta$$

where v_x and v_y are the x and y *components* of the vector.

1. An object has a velocity $v_x = 4.0$ m/s and $v_y = -3.0$ m/s. Find the speed and angle.

$$v = \sqrt{v_x^2 + v_y^2} = \sqrt{4^2 + 3^2} = \sqrt{25} = \textbf{5.0 m/s}$$

$\tan\theta = v_y/v_x = -0.75$

$\theta = \tan^{-1}(-0.75) = -37°$, or **37° below horizontal**

2. An object travels 50 m/s at an angle 60° below horizontal. Find the horizontal component of the velocity vector.

$$v_x = v \cos \theta = (50) \cos(-60°) = \textbf{25 m/s}$$

QUIZ 2.1

1. An object is initially at the origin and has a constant velocity. After 15 s, it is at (30 m, −60 m). What is the velocity?
2. An object has a velocity $v_x = 12$ m/s and $v_y = 5.0$ m/s. Find the speed v and angle θ.
3. An object travels at an angle 59° above horizontal at a speed 88 m/s. Find the vertical component of the velocity vector.
4. A rock falls from a cliff with a speed of 10 m/s. The horizontal component of the velocity is $v_x = 5$ m/s. Find the velocity's direction. Express as degrees clockwise from vertical.

ADDING VECTORS

To add two vectors, add the x components and the y components separately. For example, let vectors **A** and **B** be

$$A_x = 7, A_y = -5$$

$$B_x = -3, B_y = -1$$

The sum of these vectors is **C = A + B**,

$$C_x = 4, C_y = -6$$

1. Two vectors with magnitudes 2.0 and 4.0 are shown in Figure 2.3. Calculate their sum.

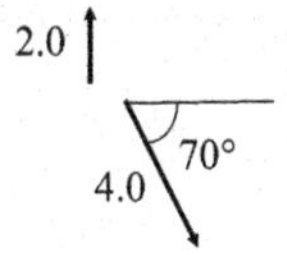

FIGURE 2.3   Two vectors.

$A_x = 0$

$A_y = 2.0$

$B_x = 4.0 \cos (70°) = 1.4$

$B_y = -4.0 \sin (70°) = -3.8$ (– because the vector points downward)

$C_x = 1.4, C_y = -1.8$

2. A boat travels north at 3.0 m/s in still water (Figure 2.4). It encounters a region where the water flows east at 0.7 m/s. Calculate the boat's velocity. Express as speed and angle with respect to north.

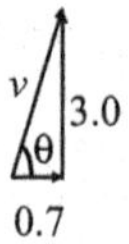

FIGURE 2.4   Addition of two vectors.

Add 3.0 m/s north to 0.7 m/s east

$v_x = 0 + 0.7 = 0.7$ m/s

$v_y = 3.0 + 0 = 3.0$ m/s

$v = \sqrt{0.7^2 + 3^2} = $ **3.1 m/s**

$\tan \theta = 3.0/0.7 = 4.3$

$\theta = \tan^{-1}(4.3) = 77°$, which is **13° east of north**

EQUATIONS OF MOTION

In Chapter 1, we used equations to calculate the velocity and position of an object. For two-dimensional motion, there are two sets of equations, one for x and one for y. We write a subscript x or y to tell them apart.

$$v_x = v_{0x} + a_x t$$

$$x = x_0 + v_{0x}t + \tfrac{1}{2}a_x t^2$$

$$v_y = v_{0y} + a_y t$$

$$y = y_0 + v_{0y}t + \tfrac{1}{2}a_y t^2$$

In these equations,

(x, y) is the position of the object at time t

(v_x, v_y) is the velocity at time t

(x_0, y_0) is the initial position $(t = 0)$

(v_{0x}, v_{0y}) is the initial velocity $(t = 0)$

(a_x, a_y) is the acceleration

1. An object is initially at the origin. It has a constant velocity $v_{0x} = 4.0$ m/s, $v_{0y} = 7.0$ m/s. Find the position after 8.0 s.

$$x_0 = y_0 = 0$$

$$a_x = a_y = 0 \text{ (constant velocity)}$$

$$x = 0 + (4)(8) + \tfrac{1}{2}0(8^2) = \mathbf{32\ m}$$

$$y = 0 + (7)(8) + \tfrac{1}{2}0(8^2) = \mathbf{56\ m}$$

2. An object is initially at the origin. It has an initial velocity $v_{0x} = 4.0$ m/s, $v_{0y} = 7.0$ m/s. It accelerates downward at $a_y = -20$ m/s². Find the position after 8.0 s.

$$a_x = 0, a_y = -20$$

$$x = 0 + (4)(8) + \tfrac{1}{2}0(8^2) = \mathbf{32\ m}$$

$$y = 0 + (7)(8) + \tfrac{1}{2}(-20)(8^2) = \mathbf{-580\ m}$$

QUIZ 2.2

1. Two vectors with magnitudes 4.0 and 1.0 are shown in Figure 2.5. Calculate their sum.

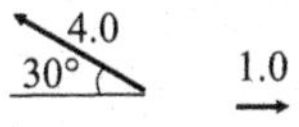

FIGURE 2.5   Two vectors.

2. An airplane points east. In still air, its speed would be 50 m/s. However, the wind blows south at 18 m/s. Calculate the airplane's velocity. Express as speed and angle with respect to east.

3. An object is initially at (–0.70 m, 0). It has a constant velocity v_{0x} = –0.50 m/s, v_{0y} = 1.0 m/s. Find the position after 6.0 s.
4. An object is initially at the origin. It has an initial velocity v_{0x} = 1.0 m/s, v_{0y} = 4.0 m/s. It accelerates toward the left at a_x = –8 m/s². Find the position after 10 s.

LAUNCHING A PROJECTILE

Figure 2.6 shows an example of projectile motion. An object is launched with an initial velocity that is horizontal. That means v_{0y} = 0.

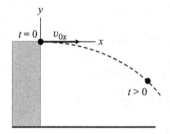

FIGURE 2.6   Object launched with an initial velocity that is horizontal.

Gravity causes the object to accelerate downward, with a_y = –10 m/s². Gravity does *not* cause acceleration in the horizontal direction, so a_x = 0.

We choose the axes so the object's initial position (t = 0) is at the origin: $x_0 = y_0 = 0$.

1. An object is launched horizontally from a 20-m tall building. The initial speed is 40 m/s. How long does it take for the object to hit the ground?

$$y = y_0 + v_{0y}t + \tfrac{1}{2}a_y t^2$$

When the object hits the ground, y = –20 m.

$$-20 = 0 + 0t + \tfrac{1}{2}(-9.8)t^2$$

$$-20 = -4.9t^2$$

$$4.1 = t^2, t = \mathbf{2.0\ s}$$

2. In the previous problem, what is the velocity just before the object hits the ground?

$v_x = v_{0x} + a_x t = 40 + (0)(2) = $ **40 m/s**

$v_y = v_{0y} + a_y t = 0 + (-9.8)(2) = $ **–20 m/s**

HITTING A TARGET

In this example, a gun is aimed at an angle θ above horizontal (Figure 2.7). It fires a projectile at a target that is some distance away. Because of gravity, the projectile will accelerate downward.

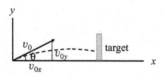

FIGURE 2.7   A projectile hits a target. The dashed line shows its trajectory.

1. A cannon shoots a ball at a wall 200 m away. The ball has an initial speed 50 m/s and angle $\theta = 60°$. How long does it take for the ball to hit the wall?

$v_{0x} = v_0 \cos \theta = (50)(\cos 60°) = 25$ m/s

$x = x_0 + v_{0x}t + \frac{1}{2}a_x t^2$

When the ball hits the wall, $x = 200$ m

$200 = 0 + (25)t + \frac{1}{2}0t^2$

$200 = 25t$, so $t = $ **8.0 s**

2. In the previous problem, at what height (y) did the ball hit the wall?

$v_{0y} = v_0 \sin \theta = (50)(\sin 60°) = 43.3$ m/s

$y = y_0 + v_{0y}t + \frac{1}{2}a_y t^2$

$y = 0 + (43.3)(8) + \frac{1}{2}(-9.8)(8^2) = $ **33 m**

QUIZ 2.3

1. An object is launched horizontally from an 80-m tall building. The initial speed is 34 m/s. How long does it take for the object to hit the ground?
2. In the previous problem, what is the velocity just before the object hits the ground?
3. A sniper aims a rifle at $1.0°$ above horizontal and fires a bullet with an initial speed 800 m/s. The target is a horizontal distance $x = 1600$ m away. How long does it take for the bullet to hit the target?
4. In the previous problem, what is the vertical position of the bullet (y) when it hits the target?

CHAPTER SUMMARY

Vector components	
	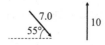
	$v_{0x} = v_0 \cos \theta$
	$v_{0y} = v_0 \sin \theta$
Equations of motion	$v_x = v_{0x} + a_x t$
	$x = x_0 + v_{0x} t + \frac{1}{2} a_x t^2$
	$v_y = v_{0y} + a_y t$
	$y = y_0 + v_{0y} t + \frac{1}{2} a_y t^2$

END-OF-CHAPTER QUESTIONS

1. A person shoots a rifle at a $76°$ angle with respect to the flat ground. The initial speed of the bullet is 100 m/s. What is the initial velocity (v_{0x} and v_{0y})?
2. Two vectors with magnitudes 7.0 and 10 are shown in Figure 2.8. Calculate their sum.

7.0
55°
10

FIGURE 2.8   Two vectors.

3. A college student throws a water balloon horizontally through an open window with a speed 7.0 m/s. The window is 45 m above the ground.
   a. How long does it take for the balloon to hit the ground?
   b. What is the velocity of the balloon just before it hits the ground?

4. A catapult launches a boulder at a 45° angle and initial speed 28 m/s. The ground is flat.
   a. How long does it take for the boulder to hit the ground?
   b. How far does the boulder travel?
5. A person throws a dart 30° above horizontal with initial speed 16 m/s. The dart hits a tree 7.0 m away.
   a. How long was the dart in the air?
   b. The dart sticks in the tree some distance above the dart's initial height. What is that distance?

ADDITIONAL PROBLEMS

1. A person is on the roof of a building 10 m above the ground. He throws a water balloon vertically upward, with initial velocity 5.0 m/s. How long does it take for the balloon to hit the ground?
2. An athlete throws a javelin at a 45° angle and initial speed 14 m/s. How long does it take for the javelin to reach its maximum height?
3. A cannon is on a hill that is 100 m above the flat ground. It shoots a cannonball at an angle 30° above horizontal and initial speed 10 m/s.
   a. How long does it take for the ball to hit the ground?
   b. What horizontal distance (x) did the ball travel?
4. A biologist aims a rifle at 60° above horizontal and shoots a tranquilizer dart with an initial speed 10 m/s. A motionless monkey is in a tree, a horizontal distance $x = 5.0$ m away. The dart hits the monkey. What was the height of the monkey (above the rifle) when it was hit?

Forces

FORCE CAUSES ACCELERATION

Force is a vector that pushes or pulls on an object. It has units of kg·m/s², which are also called newtons (N). A horizontal force acts along the x direction. If the force acts toward the right (+x direction), it's positive. If it acts toward the left, it's negative. A vertical force acts along the y direction. Up is positive, and down is negative.

Forces cause objects to accelerate. Let F_x = the sum of forces along the x direction. Force and acceleration are related by

$$F_x = ma_x$$

Similarly, letting F_y = the sum of forces along the y direction,

$$F_y = ma_y$$

1. Find the *net force* (sum of the forces) on the object in Figure 3.1.

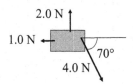

FIGURE 3.1   Three forces acting on an object.

Components of the 4.0 N vector:

$$F_x = 4 \cos (70°) = 1.37 \text{ N } (+ \text{ because the vector}$$
$$\text{points right)}$$

$F_y = -4 \sin (70°) = -3.76$ N ($-$ because the vector points down)

Add the components:

$F_x = -1 + 1.37 = \mathbf{0.37}$ **N**

$F_y = 2 - 3.76 = \mathbf{-1.76}$ **N**

2. A rocket in outer space has a mass of 1000 kg. Its engine produces a thrust of 20,000 N. What is the rocket's acceleration?

$F_x = ma_x$

$20,000 \text{ N} = (1000 \text{ kg}) \, a_x$

$\mathbf{20 \text{ m/s}^2} = a_x$

GRAVITY

The force of gravity acts vertically downward (toward the center of the earth). If the y axis points straight up, then the gravitational force is

$$F_y = -mg$$

where m is the object's mass and $g = 9.8$ m/s². Gravity does not act along the horizontal direction. The magnitude of the force is called the object's *weight*. Weight is a positive number.

1. A person has a mass of 100 kg. What is the gravitational force on the person and the person's weight?

$F_y = -mg$

$= -(100)(9.8)$

$= \mathbf{-980}$ **N**

Weight $= \mathbf{980}$ **N**

(Note that kg is a unit of mass, not weight)

FIGURE 3.2   Object in freefall.

2. An object is in freefall (Figure 3.2). What is its acceleration?

$$F_y = ma_y$$

$$-mg = ma_y$$

$$-g = a_y$$

$$-9.8 \text{ m/s}^2 = a_y$$

Notice how the mass canceled. Any object in freefall has the same acceleration, regardless of mass!

QUIZ 3.1

1. Find the net force on the object in Figure 3.3.

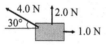

FIGURE 3.3   Three forces acting on an object.

2. A car's engine produces a horizontal force of 800 N. The car mass is 1000 kg. What is the acceleration?
3. A block has a mass of 8.0 kg. What is the gravitational force on the block? What is the block's weight?
4. Galileo drops two masses, 1.0 kg and 8.0 kg, from the Leaning Tower of Pisa. What is the acceleration of each mass?

FORCE BETWEEN TWO OBJECTS

Suppose a person pulls a crate in the positive x direction with a force of 30 N (Figure 3.4).

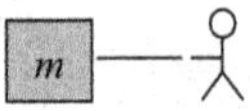

FIGURE 3.4   Person pulling a crate.

Look at the force on each object (Figure 3.5). The crate experiences a force $F_x = +30$ N, and the person experiences an *equal but opposite* force $F_x = -30$ N.

Crate: Person:

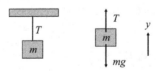

FIGURE 3.5   Force diagrams showing the force exerted on the crate and on the person.

1. A truck pulls a trailer. The truck exerts a force $F_x = -700$ N on the trailer. What force does the trailer exert on the truck?

$F_x = 700$ **N**

2. A person pushes an elephant with a force $F_x = 20$ N. What force does the elephant exert on the person?

$F_x = -20$ **N**

TENSION

When a rope pulls on an object, the rope gets taut or stretched. A rope that is stretched is under *tension* (T). T is the amount of force that the rope exerts on the object.

Ropes (or strings, chains, cables, etc.) only pull. They never push.

1. A 5.0-kg mass hangs from the ceiling by a rope (Figure 3.6). What is the tension of the rope?

FIGURE 3.6   A mass hanging by a rope. The force diagram is on the right.

From the force diagram on the right, $F_y = T - mg$
The object is not accelerating, so $F_y = ma_y = 0$

$T - mg = 0$

$T = mg$

$T = (5)(9.8) = \mathbf{49\ N}$

2. A string pulls horizontally on a cat toy (Figure 3.7). The mass of the toy is 0.10 kg and its acceleration is 2.0 m/s². What is the tension of the string?

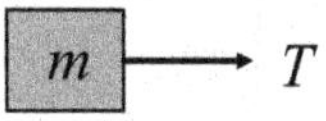

FIGURE 3.7 Tension force acting on an object.

From the force diagram, $F_x = T$

$F_x = ma_x = (0.1)(2) = 0.2\ N$

$T = \mathbf{0.20\ N}$

QUIZ 3.2

1. A child pulls a wagon in the $-x$ direction. The child pulls with a force $F_x = -5.0\ N$. What force does the wagon exert on the child?
2. An athlete pushes a basketball upward with a force 10 N. What is the force that the basketball exerts on the athlete?
3. A wrecking ball with a mass of 300 kg is attached to a crane by a cable. The ball is motionless. What is the tension in the cable?
4. In a toy train, a caboose with a mass of 0.10 kg is pulled horizontally by a string. The acceleration is 0.50 m/s². What is the tension in the string?

NORMAL FORCE

An object sits on a horizontal surface (Figure 3.8). The force of gravity acts downward. Why doesn't the object accelerate downward?

FIGURE 3.8 Object on a horizontal surface.

The reason is that the surface exerts a force on the object. The direction of this force is perpendicular, or *normal*, to the surface. We label this force *n*.

Figure 3.9 is a force diagram for the object. Notice that the *n* vector is perpendicular to the surface, and points away from the surface.

FIGURE 3.9 Force diagram for an object on a horizontal surface.

1. An object with a mass of 6.0 kg sits on a table. What is the normal force on the object?

   From the force diagram, $F_y = n - mg$
   The object is not accelerating, so $F_y = ma_y = 0$

   $n - mg = 0$

   $n = mg$

   $n = (6)(9.8) = \textbf{59 N}$

2. An astronaut with a mass of 100 kg sits on a seat in a rocket. The rocket accelerates upward at 10 m/s². What is the normal force exerted by the seat?

   From the diagram, $F_y = n - mg$
   The astronaut is accelerating. $F_y = ma_y = (100)(10) = 1000$ N

   $n - mg = 1000$

   $n = 1000 + mg$

   $n = 1000 + (100)(9.8) = \textbf{1980 N}$

FRICTION

Consider a block on a table. The surface of the block is in contact with the surface of the table. These surfaces are not perfectly smooth, so it requires effort to make the block slide. The roughness of the surfaces causes a *friction* (f) force. The friction force vector is parallel to the surfaces. Friction opposes sliding. It never causes sliding.

Static friction (f_s) prevents sliding motion. It occurs when the objects are not moving relative to each other (no sliding). The maximum magnitude of static friction is

$$f_s(\text{max}) = \mu_s n$$

where μ_s is the *coefficient of static friction*. It depends on the materials (e.g., wood, metal, etc.). A large normal force n produces a large friction force. If a force exceeds $f_s(\text{max})$, the block will begin to slide. When sliding occurs, we have *kinetic friction* (f_k),

$$f_k = \mu_k n$$

where μ_k is the *coefficient of kinetic friction*.

1. A wood block ($m = 3.0$ kg) sits on a wood table (Figure 3.10). A horizontal force F pushes on the block. How much force is required to make the block slide? Let $\mu_s = 0.20$ and $\mu_k = 0.10$.

FIGURE 3.10   Object on a table. The force diagram is on the right.

The object is not accelerating vertically, so $n = mg$

$n = (3)(9.8) = 29.4$ N

$f_s(\text{max}) = (0.2)(29.4) = 5.9$ N

This is the maximum static friction force. To overcome it, $F > $ **5.9 N**

2. In the previous problem, let $F = 12$ N. Find the block's acceleration.

Because 12 N exceeds f_s(max), the block is sliding.

$f_k = (0.1)(29.4) = 2.94$ N

$F_x = F - f_k = 12 - 2.94 = 9.1$ N

$F_x = ma_x$

$9.1 = 3a_x$, $a_x = \textbf{3.0 m/s}^2$

QUIZ 3.3

1. A book has a mass of 1.0 kg. It sits on a shelf. What is the normal force on the book?
2. On a carnival ride, an 80-kg person sits on a chair that accelerates downward, $a_y = -5.0$ m/s^2. What is the normal force on the person?
3. A person pushes a full cardboard box with a mass of 100 kg. How much force is required to make the box move? Let $\mu_s = 0.30$.
4. In the previous problem, suppose the person pushes with a 500 N horizontal force. What is the acceleration of the box? Let $\mu_k = 0.25$.

CHAPTER SUMMARY

Force causes acceleration	$F_x = ma_x$ $F_y = ma_y$
Force between two objects	If A exerts a force on B, then B exerts an equal but opposite force on A.
Gravitational force	$F_y = -mg$ (if y axis points straight up)
Tension (T)	A rope pulls with a force T
Normal force (n)	Perpendicular to surface
Friction force (f)	Parallel to surface, opposes sliding
Static friction	f_s(max) $= \mu_s n$
Kinetic friction	$f_k = \mu_k n$

END-OF-CHAPTER QUESTIONS

1. Four forces act on an object with mass, m, equal to 4.0 kg (Figure 3.11). What is the acceleration?

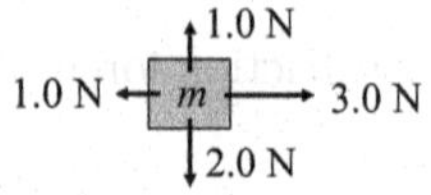

FIGURE 3.11   Four forces acting on an object.

2. A large truck collides with a small car. The truck exerts a force $F_x = 10{,}300$ N on the car. What force does the car exert on the truck?

3. A 0.53-kg salmon is on a plate. What is the normal force on the salmon?

4. A 1.0-kg mass is attached to a string. A person pulls the string upward, causing the mass to accelerate at 5.0 m/s². What is the tension in the string?

5. A horse pulls a 300-kg mass horizontally. The coefficient of static friction is 0.5. The coefficient of kinetic friction is 0.3.
   a. How much force does the horse need to exert to move the mass?
   b. Suppose the horse pulls with 2000 N of force. What is the acceleration of the mass?

ADDITIONAL PROBLEMS

1. A race car accelerates with $a_x = 10$ m/s². The driver has a mass of 100 kg. Ignore friction.
   a. What is the normal force of the seat bottom on the diver? Assume the seat bottom is a horizontal surface.
   b. What is the normal force of the seat back on the driver? Assume the seat back is a vertical surface.

2. A rope and chain are attached to an 8.0-kg mass. The rope pulls toward the left with a force of 12 N. The chain pulls toward the right. The mass accelerates toward the right at 16 m/s². What is the tension in the chain? Ignore friction.

3. A person throws a disk onto a smooth surface. The disk has an initial horizontal velocity of 3.0 m/s. The coefficient of kinetic friction is 0.050. How long does it take for the disk to stop?

4. A cart accelerates toward the right. A book with a mass of 0.50 kg sits on the cart and does not slide. The coefficient of static friction is 0.30.
   a. Suppose the acceleration is 2.0 m/s². What is the friction force on the book?
   b. What is the maximum acceleration such that the book will not slide?

Applying Newton's Laws

4

TENSION

$\mathbf{F} = m\mathbf{a}$ is Newton's second law. Newton's first law is a special case of this; it says that if $\mathbf{F} = 0$, $\mathbf{a} = 0$ (i.e., the velocity is constant if there is no net force). Newton's third law states that if A exerts a force on B, then B exerts an equal but opposite force on A.

Force is a vector, with components shown in Figure 4.1.

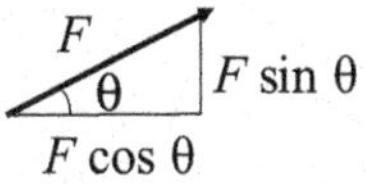

$$F \quad \theta \quad F \sin \theta$$
$$F \cos \theta$$

FIGURE 4.1 Components of a force vector.

In this chapter, we discuss several examples that use Newton's laws, starting with tension.

1. An object is suspended by two identical ropes (Figure 4.2). What is the tension in each rope?

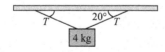

FIGURE 4.2 Object suspended by two ropes.

For one rope, $F_y = T \sin (20°) = 0.34\ T$

For both ropes, $F_y = 2 \times 0.34\ T = 0.68\ T$

Force of gravity: $F_y = -mg = -(4)(9.8) = -39$ N

$F_y(\text{total}) = 0.68\ T - 39 = 0$

$0.68\ T = 39,\ T = \mathbf{57\ N}$

2. A child pulls a 10-kg sled by a rope, causing an acceleration, $a_x = -2.0$ m/s^2 (Figure 4.3). What is the tension in the rope? Ignore friction.

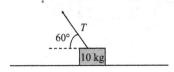

FIGURE 4.3   Tension acting on a 10-kg object.

$F_x = -T\cos(60°) = -0.5\ T\ (-\text{ because the vector points left})$

$F_x = ma_x$

$-0.5\ T = (10)(-2)$

$T = 20/0.5 = \mathbf{40\ N}$

NORMAL FORCE

Here are examples of calculating the normal force, when another force acts at some angle. In these examples, the object is not accelerating in the y direction. Therefore, the sum of the forces in the y direction must be 0.

1. A person pushes a cart with a mass of 20 kg (Figure 4.4). The person exerts a force 100 N at 55° with respect to horizontal. What is the normal force n of the ground on the cart?

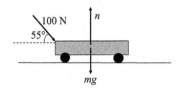

FIGURE 4.4   Forces acting on a cart.

Person's force: $F_y = -100\sin(55°) = -82\ N$

$F_y(\text{total}) = n - mg - 82 = 0$

$n = mg + 82$

$n = (20)(9.8) + 82 = \mathbf{278\ N}$

2. A rope pulls a 50-kg block at an angle 30° above horizontal (Figure 4.5). The tension of the rope is 200 N. What is the friction force? The coefficient of kinetic friction is 0.10.

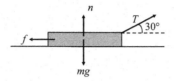

FIGURE 4.5   Forces acting on a block.

Rope: $F_y = 200 \sin (30°) = 100$ N

F_y (total) $= n - mg + 100 = 0$

$n = mg - 100 = 390$ N

$f = \mu_k n = (0.1)(390) = \textbf{39 N}$

QUIZ 4.1

1. A 0.14-kg object is suspended by two wires. Each wire is at a horizontal angle 1.0°. Find the tension in each wire.
2. A 2.0-kg object is on a frictionless floor. A person pushes on the object at a 30° angle with respect to the ground. The person exerts a force 4.6 N. Find the acceleration of the object.
3. In the previous problem, what is the normal force on the object?
4. A rope pulls on a block (Figure 4.6). The coefficient of kinetic friction is 0.050. Find the friction force (magnitude and direction).

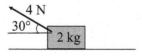

FIGURE 4.6   Rope pulling on a block.

INCLINED PLANE WITHOUT FRICTION

A block is on a frictionless plane, which is inclined at an angle θ above horizontal (Figure 4.7). We choose axes where x is parallel to the surface, and y is perpendicular. Imagine slowly increasing θ from 0. The axes tilt by an angle θ. Notice how the y axis makes an angle θ with vertical.

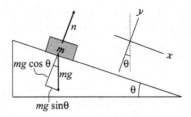

FIGURE 4.7   Block on an inclined plane.

The normal force n points in the y direction. However, we must find the x and y components of mg. The x component is $mg \sin \theta$. The y component is $-mg \cos \theta$ (− because it points down).

The mass slides on the surface, accelerating in the x direction. It does *not* accelerate in the y direction because y is perpendicular to the surface.

1. What is the acceleration of a 4.0-kg mass on a frictionless inclined plane with $\theta = 30°$?

   From the diagram, $F_x = mg \sin \theta$

   $F_x = ma_x$

   $ma_x = mg \sin \theta$

   $a_x = g \sin \theta$

   $a_x = 9.8 \sin (30°) = \textbf{4.9 m/s}^2$

2. In the previous problem, what is the normal force?

   The mass does not accelerate in the y direction, so $F_y = 0$.

   $F_y = n - mg \cos \theta = 0$

   $n = mg \cos \theta$

   $n = (4)(9.8) \cos (30°) = \textbf{34 N}$

INCLINED PLANE WITH FRICTION

We now include the friction force f, which opposes sliding (Figure 4.8). Because the mass wants to slide in the $+x$ direction, friction points in the $-x$ direction.

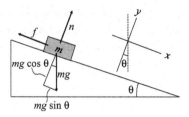

FIGURE 4.8   Block on an inclined plane with friction.

1. A block sits on an inclined plane. The angle θ is slowly increased from 0. At what angle will the block begin to slide? Let $\mu_s = 0.20$.

$n = mg \cos \theta$

$f_s(\text{max}) = \mu_s n = \mu_s mg \cos \theta$

From the diagram, $F_x = mg \sin \theta - f$

Find the angle such that $f = f_s$ (max) and $F_x = 0$

$0 = mg \sin \theta - \mu_s mg \cos \theta$

$\sin \theta = \mu_s \cos \theta$

$\tan \theta = \mu_s$

$\theta = \tan^{-1} (\mu_s) = \tan^{-1} (0.2) = \mathbf{11°}$

2. In the previous problem, if $\theta = 30°$, find the acceleration. Let $\mu_k = 0.10$.

From the diagram, $F_x = mg \sin \theta - f$

$F_x = mg \sin \theta - \mu_k mg \cos \theta$

$F_x = ma_x$

$ma_x = mg \sin \theta - \mu_k mg \cos \theta$

$a_x = g \sin \theta - \mu_k g \cos \theta$

$a_x = (9.8)(0.5) - (0.1)(9.8)(0.866) = \mathbf{4.1 \ m/s^2}$

QUIZ 4.2

1. A mass slides down a 45° frictionless incline. What is the acceleration?
2. A 3.0-kg mass is on a plane inclined at a horizontal angle 20°. Find the magnitude of the normal force.

3. A mass sits on an inclined plane. What is the minimum angle such that the mass will slide? The coefficient of static friction is 0.40.
4. A mass slides down a 60° incline. The coefficient of kinetic friction is 0.20. Find the acceleration.

SIMPLE PULLEY

In a simple pulley, a rope and disc are used to redirect a force (Figure 4.9). A person pulls on the rope with a force of magnitude F. (The rope pulls on the person with an equal but opposite force.) The tension of the rope is equal to F. If the person pulls with $F = 17$ N, then $T = 17$ N.

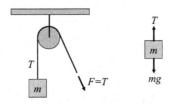

FIGURE 4.9   Simple pulley. A force diagram is on the right.

1. A 20-kg mass is connected to a simple pulley. A person pulls on the rope with 300 N of force. What is the acceleration of the mass?

From the force diagram, $F_y = T - mg$

$F_y = ma_y$

$ma_y = T - mg$

$20a_y = 300 - (20)(9.8)$

$a_y = 104/20 = \mathbf{5.2 \ m/s^2}$

2. In the previous problem, what is the minimum amount of force required to lift the mass?

$F_y = T - mg$

Find the point where F_y is exactly 0.

$0 = T - mg$

$T = mg = \mathbf{196 \ N}$

ATWOOD'S MACHINE

In Atwood's machine, two masses are connected to each other via a simple pulley (Figure 4.10). The lighter mass m accelerates up with $a_y = a$. The heavier mass, M, accelerates down with $a_y = -a$. To find the acceleration, first solve for T for each mass. Then set the two equations equal to each other and solve for a.

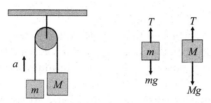

FIGURE 4.10   Atwood's machine. The force diagrams are on the right.

1. An Atwood's machine has masses $m = 10$ kg and $M = 30$ kg. Find the acceleration of mass m.

   Mass m:

   $F_y = T - mg$

   $F_y = ma$

   $T - mg = ma$

   Indicate $T = mg + ma = 98 + 10a$ as (1)

   Mass M:

   $F_y = T - Mg$

   $F_y = M(-a)$

   $T - Mg = -Ma$

   Indicate $T = Mg - Ma = 294 - 30a$ as (2)

   Set the underlined equations (1) and (2) equal to each other:

   $98 + 10a = 294 - 30a$

   $40a = 196$

   $a = \mathbf{4.9\ m/s^2}$

2. In the previous problem, what is the tension in the rope?

From the first underlined equation (1):

$T = 98 + 10a$

$T = 98 + 490 = \textbf{588 N}$

QUIZ 4.3

1. A 3.0-kg mass is attached to a simple pulley. A person exerts a gentle force of only 6.0 N on the rope. What is the acceleration of the mass?
2. In the previous problem, what is the minimum force required to lift the mass?
3. An Atwood's machine has masses 10 kg and 40 kg. What is the acceleration of the heavier mass?
4. In the previous problem, what is the tension in the rope?

CHAPTER SUMMARY

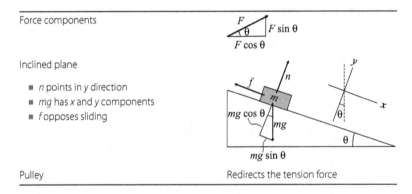

Force components	F, $F \sin \theta$, $F \cos \theta$
Inclined plane • n points in y direction • mg has x and y components • f opposes sliding	
Pulley	Redirects the tension force

END-OF-CHAPTER QUESTIONS

1. A mass $m = 2.0$ kg is suspended by two cables (Figure 4.11). What is the tension in each cable?

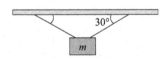

FIGURE 4.11   Mass suspended by two cables.

2. A person pulls a 100-kg object with 30 N of force at 50° above horizontal. The object is on a frictionless horizontal surface.
   a. What is the acceleration of the object?
   b. What is the normal force on the object?
3. A block slides down a plane inclined at 30°.
   a. What is the acceleration if the plane is frictionless?
   b. Find the acceleration if the plane has a coefficient of kinetic friction of 0.25.
4. A 22-kg block sits motionless on a plane inclined at 27°. What is the magnitude of static friction force on the block?
5. An Atwood's machine has two masses that are 5.0 kg each.
   a. What is the tension in the rope?
   b. An additional 5.0 kg is added to one of the masses. What is the acceleration of the lighter mass?

ADDITIONAL PROBLEMS

1. A person pushes down on a 40-kg cart at a 53° angle with respect to horizontal, with 100 N of force. The cart is on a horizontal, frictionless surface.
   a. What is the normal force on the cart?
   b. The cart starts from rest. What is the displacement after 4.0 s?
2. Find the acceleration of the 10-kg mass in Figure 4.12 where (a) $T = 100$ N, and (b) $T = 200$ N. Ignore friction.

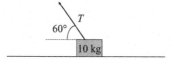

FIGURE 4.12   Tension force pulling on a mass.

3. A worker wants to push a 100-kg crate up a 30° slope. The coefficient of static friction is 0.80. What is the minimum force (magnitude) the worker must exert?
4. Two masses, $m = 5.0$ kg and $M = 15$ kg, are connected via a pulley (Figure 4.13). For M, the coefficient of kinetic friction is 0.10. The masses are accelerating. What is the acceleration of M?

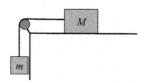

FIGURE 4.13   Two masses connected via a pulley.

Circular Motion and Gravity

5

CENTRIPETAL ACCELERATION

In *uniform circular motion*, an object travels in a circle of radius, R, with a constant speed, v (Figure 5.1). The velocity is *tangential* to the circle. Even though the speed doesn't change, the direction of the velocity vector continuously changes as the object goes around. This is a type of acceleration because acceleration is change in velocity over time. We call it *centripetal acceleration*, where "centripetal" means pointing toward the center.

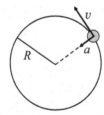

FIGURE 5.1 Object traveling in a circle.

Follow these rules to calculate centripetal acceleration:

- The acceleration vector points toward the center of the circle.
- The magnitude of the acceleration is $a = v^2/R$.

1. An object travels in a circular path with a speed 5.0 m/s. The radius of the circle is 10 m. Find the acceleration.

$a = v^2/R$

$a = 25/10 = \mathbf{2.5\ m/s^2}$

Direction: toward center of circle

2. A bug is on a disc that rotates at 77 rotations per minute (rpm). The distance from the bug to the disc center is 10 cm. What is the bug's acceleration?

Circumference $= 2\pi R = 0.63$ m

$$v = \frac{77\,\text{rotations}}{\text{min}}\,\frac{1\,\text{min}}{60\,\text{s}}\,\frac{0.63\,\text{m}}{\text{rotation}} = 0.81\,\text{m/s}$$

$a = v^2/R$

$a = 0.81^2/0.1 = \mathbf{6.6\ m/s^2}$

Direction: toward center of circle

MOTION IN A HORIZONTAL CIRCLE

Consider an object attached to a string on a frictionless table (Figure 5.2). The object undergoes uniform circular motion. We define the y direction to point from the center to the object. The tension force, T, points in the $-y$ direction. It pulls the ball toward the center and is referred to as a *centripetal force*. (If someone cuts the string, the ball will go in a straight line.)

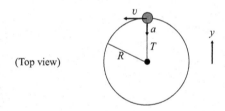

(Top view)

FIGURE 5.2   Object attached to a string and undergoing uniform circular motion.

1. A 0.20-kg ball attached to a string is on a frictionless table. The ball travels in a horizontal circular path with a speed 5.0 m/s. The radius of the circle is 0.10 m. Find the tension of the string.

$a_y = -v^2/R$

$F_y = -T$

$F_y = ma_y$

$-T = -mv^2/R$

$T = (0.2)(25)/0.1 = \mathbf{50\ N}$

2. In the previous problem, what is the normal force of the table on the ball (Figure 5.3)?

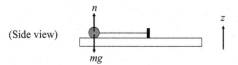

(Side view)

FIGURE 5.3 Side view of Figure 5.2.

$$F_z = n - mg = 0$$

$$n = mg$$

$$n = (0.2)(9.8) = \textbf{2.0 N}$$

QUIZ 5.1

1. An object travels in a circular path with a speed 3.0 m/s. The radius of the circle is 9.0 m. Find the acceleration.
2. A small rock is stuck on a car tire. The tire has a diameter of 75 cm and spins at 1200 rpm. What is the acceleration of the rock?
3. A 1.0-kg mass is attached to a rope on a horizontal frictionless surface. The mass travels in a horizontal circle of radius 0.50 m with a constant speed 8.0 m/s. Calculate the tension of the rope.
4. In the previous problem, what is the normal force on the mass?

CENTRIPETAL FRICTION FORCE

A car traveling around a curve requires the centripetal force of friction. Otherwise, it will travel in a straight line and go off the road. This can happen when the road is icy. In Figure 5.4, a car is going around a circular horizontal track at a speed, v. The friction force, f, acts in the $-y$ direction. Because f prevents sliding, it is considered static friction, even though the car is moving.

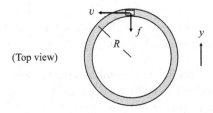

(Top view)

FIGURE 5.4 Car traveling around a horizontal track.

1. A 1500-kg car travels around a horizontal circular track of radius 500 m at a speed 20 m/s. What is the centripetal friction force?

$a_y = -v^2/R$

$F_y = -f$

$F_y = ma_y$

$-f = -mv^2/R$

$f = (1500)(400)/500 = \mathbf{1200\ N}$

2. In the previous problem, the coefficient of static friction is 0.50. What is the maximum speed the car can travel without sliding?

$f(\text{max}) = \mu_s mg$

$f(\text{max}) = mv^2/R$

$\mu_s mg = mv^2/R$

$\mu_s gR = v^2$

$v^2 = (0.5)(9.8)(500) = 2450$

$v = \sqrt{2450} = \mathbf{49\ m/s}$

MOTION IN A VERTICAL CIRCLE

The previous examples involved motion in a horizontal circle. Now suppose an object travels in a vertical circle (Figure 5.5). Defining z to point up, gravity acts in the −z direction.

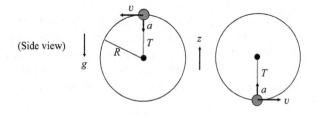

FIGURE 5.5 Object traveling in a vertical circle. The object is shown at two different times, at the top and bottom of the circular path.

1. A 0.20-kg ball is attached to a motor by a 0.50-m long massless rod. The motor makes the ball travel in a vertical circle with a speed $v = 3.0$ m/s. What is the tension of the rod when the ball is at the top of the circle?

$a_z = -v^2/R$

$F_z = -T - mg$

$F_z = ma_z$

$T + mg = mv^2/R$

$T + (0.2)(9.8) = (0.2)(9)/0.5$

$T = 3.6 - 1.96 = \mathbf{1.6\ N}$

2. In the previous problem, what is the tension when the ball is at the bottom of the circle?

$a_z = v^2/R$

$F_z = T - mg$

$F_z = ma_z$

$T - mg = mv^2/R$

$T - (0.2)(9.8) = (0.2)(9)/0.5$

$T = 3.6 + 1.96 = \mathbf{5.6\ N}$

QUIZ 5.2

1. A 100-kg go-cart travels around a horizontal circular track of radius 50 m at a speed 5.0 m/s. The coefficient of static friction is 0.20. What is the centripetal friction force?
2. In the previous problem, what is the maximum speed that the go-cart can go and not slide?
3. A 3.0-kg mass is held by a rope. A person makes it travel in a vertical circle of diameter 1.0 m. At the top of the circle, the speed of the mass is 4.0 m/s. What is the tension of the rope?
4. In the previous problem, at the bottom of the circle, the speed is 6.0 m/s. What is the tension of the rope?

NEWTON'S LAW OF GRAVITY

A mass, m, experiences a gravitational force due to mass, M. The direction is toward M (i.e., it is an attractive force). The magnitude is given by Newton's law of gravity,

$$F = \frac{GmM}{r^2}$$

where everything is in SI units, $G = 6.67 \times 10^{-11}$, and r is the distance between the objects' centers.

Suppose a mass, m, is on the surface of a planet of radius, R. The force of gravity on this mass is $F = mg$. Equating this with Newton's law of gravity,

$$g = \frac{GM}{R^2}$$

Plugging in the values for Earth ($M = 6.0 \times 10^{24}$ kg, $R = 6.4 \times 10^6$ m), we get $g = 9.8$ m/s². Different planets or moons have different g values.

1. Find the magnitude of the gravitational force on a 1000-kg object due to a 10,000-kg object that is 10 m away.

$$F = (6.67 \times 10^{-11})(1000)(10,000)/100 = \mathbf{6.7 \times 10^{-6}\ N}$$

2. By how much does g decrease when you go to the top of an 828-m tall building?

At the surface, $g = (6.67 \times 10^{-11})(6.0 \times 10^{24})/(6,400,000)^2 = 9.77051$

At the top, $g = (6.67 \times 10^{-11})(6.0 \times 10^{24})/(6,400,828)^2 = 9.76798$

Difference $= 9.77051 - 9.76798 = \mathbf{2.5 \times 10^{-3}\ m/s^2}$

(Note that the calculated g values are not highly accurate, but the difference between them is.)

CIRCULAR ORBITS

A satellite going around a planet has a (nearly) circular orbit (Figure 5.6). From $F = ma$, we have

$$\frac{GmM}{r^2} = m\left(\frac{v^2}{r}\right)$$

Solving for the satellite's speed yields

$$v = \sqrt{\frac{GM}{r}}$$

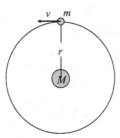

FIGURE 5.6 Satellite orbiting a planet.

1. Calculate the speed of a satellite that orbits Earth at an altitude of 500 km.

The radius of the earth is 6.4×10^6 m

$r = 6.4 \times 10^6 + 5 \times 10^5 = 6.9 \times 10^6$ m

$$v = \sqrt{\frac{(6.67\times10^{-11})(6\times10^{24})}{6.9\times10^{6}}} = 7600 \text{ m/s}$$

2. Derive an expression for the period of a satellite's orbit squared (T^2).

The time it takes to complete one orbit is $T = 2\pi r/v$.

$$T = \frac{2\pi r}{\sqrt{\dfrac{GM}{r}}}$$

$$T^2 = \frac{2^2\pi^2 r^2}{\left(\dfrac{GM}{r}\right)}$$

Multiply top and bottom by r:

$$T^2 = \frac{4\pi^2 r^3}{GM}$$

QUIZ 5.3

1. Two students with masses of 65 kg and 75 kg, respectively, sit 0.80 m away from each other. Calculate the magnitude of the gravitational force.
2. Saturn has a mass of 5.7 × 10²⁶ kg and radius 5.9 × 10⁷ m. Calculate g on the surface of Saturn.
3. The Earth orbits the Sun at a radius of approximately 150 million km. The mass of the Sun is 2.0 × 10³⁰ kg. Estimate the Earth's orbital speed, assuming a circular orbit.
4. In deep space, a rock orbits a 100,000-kg asteroid at a radius of 30 m. Calculate the orbital period.

CHAPTER SUMMARY

Centripetal acceleration	$a = v^2/R$ Direction: Toward center of circle
Newton's law of gravity	$F = \dfrac{GmM}{r^2}$ $G = 6.67 \times 10^{-11}$ (SI units)
Gravitational acceleration	$g = \dfrac{GM}{R^2}$
Orbital speed	$v = \sqrt{\dfrac{GM}{r}}$
Orbital period	$T^2 = \dfrac{4\pi^2 r^3}{GM}$

END-OF-CHAPTER QUESTIONS

1. A 1.0-gram particle is on a disc that rotates at 120 rotations per minute (rpm). The distance from the particle to the disc center is 10 cm. What is the centripetal friction force on the particle?
2. A car travels around a horizontal circular track of radius 300 m. The coefficient of static friction is 0.40. What is the maximum speed the car can travel without sliding?
3. A 3.0-kg rock is attached to a 0.80-m rope. A person swings the rock in a vertical circle. At the bottom of the swing, its speed is 4.0 m/s. What is the tension of the rope at that point?
4. An 800-kg vehicle travels over the top of a hill at 35 m/s. The top of the hill is circular with a radius of 200 m. Find the normal force on the vehicle.
5. Find the speed and period for a satellite orbiting the earth at an altitude of 20,000 km.

ADDITIONAL PROBLEMS

1. In a science fiction movie, a circular space station with a radius 500 m rotates. A person standing on the edge of the space station experiences a centripetal normal force (n). Find the speed of rotation (v) that would be required to simulate Earth's gravity.
2. A toy of mass 0.64 kg is attached to a string of length 0.32 m. The toy is on a horizontal table and travels in a circle at 60 rpm. The coefficient of kinetic friction is 0.051.
   a. What is the tension in the string?
   b. To keep the speed constant, the toy has a small motor. How much force provided by the motor?
3. A 5.0-kg tether ball is held by a rope at a vertical angle 45°. The ball travels in a horizontal circle of radius 1.6 m at a speed 4.0 m/s. What is the tension in the rope?
4. At the base of a tall building, $g = 9.8000$ m/s². At the top, $g = 9.7987$ m/s². How tall is the building?

Work and Energy

WORK

You do positive work when you push or pull an object, and it moves in the same direction as the force you exert. If you push on an elephant and it doesn't move, you haven't done any work. Work (W) is measured in Joules (J). The work done by a force is

$$W = F_{//}d$$

where $F_{//}$ is the component of the force *along the direction of motion*, and d is the distance over which the force acts. If the force is opposite to the direction of motion, then $F_{//}$ is negative.

1. A person pushes a cart with $F = 5.0$ N at $\theta = 30°$ over a distance $d = 10$ m (Figure 6.1). How much work did the person do?

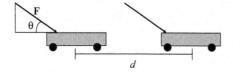

FIGURE 6.1 A force pushes a cart, which moves a distance, d.

$$W = F_{//}d$$
$$= (F \cos \theta)d$$
$$= (5 \cos 30°)(10) = \textbf{43 J}$$

2. A barbell has two masses $m = 25$ kg connected by a massless bar (Figure 6.2). A person slowly lifts the barbell a distance $h = 0.80$ m. How much work was done by the person and by gravity?

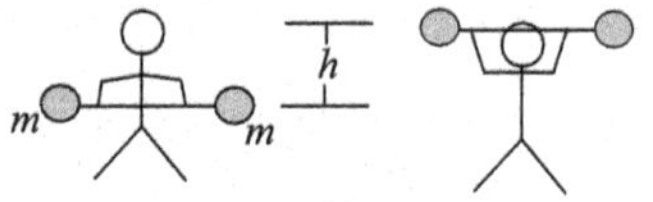

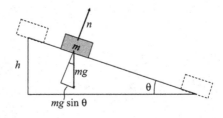

FIGURE 6.2   Person lifting a barbell of mass 2m.

The weight is $2\,mg = 2(25)(9.8) = 490$ N

$W = F_{//}d$

Person: $W = (490)(0.8) = \mathbf{390\ J}$

Gravity: $W = (-490)(0.8) = \mathbf{-390\ J}$

WORK AND AN INCLINED PLANE

Recall the inclined plane (Figure 6.3). An object is initially at a height h. It slides down the inclined plane and reaches the bottom. The distance traveled, d, is the hypotenuse of the triangle. From trigonometry,

$$\sin \theta = h/d$$

FIGURE 6.3   Object on an inclined plane. Its initial height is h.

1. How much work was done by the normal force?

$W = F_{//}d$

Because the normal force is perpendicular to the direction of motion,

$F_{//} = 0$

$\mathbf{W = 0}$

2. How much work was done by gravity?

$W = F_{||}d$

$= (mg \sin \theta)d$

$= (mg \, h/d)d = \textbf{mgh}$

QUIZ 6.1

1. A child pulls a wagon with a force of 7.0 N at an angle 60° above horizontal over a distance of 4.0 m. How much work did the child do?
2. A person lifts a 50-kg object 0.50 m. How much work did the person do?
3. In the previous problem, how much work was done by gravity?
4. A 1000-kg car is on a hill that is 50-m high. It rolls to the bottom of the hill. How much work was done by gravity?

KINETIC ENERGY

Kinetic energy (K) is due to an object's motion, and its units are Joules (J). It is defined

$$K = \tfrac{1}{2}mv^2$$

where m is the object's mass, and v is its speed. Positive work increases K. Negative work decreases K. This change in K can be written

$$K_f = K_i + W$$

where K_f and K_i are the final and initial kinetic energies.

1. A 3.0-kg block travels in a circle with a speed of 4.0 m/s. What is its kinetic energy?

$K = \tfrac{1}{2}mv^2$

$K = \tfrac{1}{2}(3)(4^2) = \textbf{24 J}$

2. A 2.0-kg block's initial speed is 3.0 m/s. A force of 8.0 N pushes the block over a distance of 2.0 m. What is the final speed?

$K_i = \tfrac{1}{2}(2)(3^2) = 9 \text{ J}$

$W = (8)(2) = 16 \text{ J}$

$K_f = 9 \text{ J} + 16 \text{ J} = 25 \text{ J}$

$K_f = \tfrac{1}{2}mv^2$

$25 = \tfrac{1}{2}(2)v^2$

$25 = v^2, v = \textbf{5.0 m/s}$

POTENTIAL ENERGY

Potential energy is stored energy that can be converted to kinetic energy. The potential energy (U) due to gravity is

$$U = mgy$$

where m is the object's mass, and y is the object's height. If there is no friction or other external force, then the total energy E is constant:

$$E = \tfrac{1}{2}mv^2 + U = \text{constant}$$

This property is called *energy conservation*. The initial energy E_i equals the final energy E_f.

1. An object is thrown straight up at 7.0 m/s. How high will it go?

$$E_i = \tfrac{1}{2}mv_i^2 + mg0 = \tfrac{1}{2}mv_i^2$$

$$E_f = \tfrac{1}{2}m0^2 + mgy = mgy$$

$$\tfrac{1}{2}mv_i^2 = mgy$$

$$\tfrac{1}{2}v_i^2/g = y$$

$$y = \tfrac{1}{2}(49)/9.8 = \textbf{2.5 m}$$

2. A roller coaster starts at a height h from rest (Figure 6.4). It rolls down and goes on a loop of radius R. What is the speed at the top of the loop?

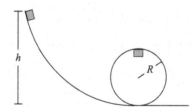

FIGURE 6.4   Roller coaster.

$$E_i = \tfrac{1}{2}m0^2 + mgh = mgh$$

$$E_f = \tfrac{1}{2}mv^2 + mg(2R)$$

$$mgh = \tfrac{1}{2}mv^2 + mg(2R)$$

$$gh = \tfrac{1}{2}v^2 + 2gR$$

$$2gh - 4gR = v^2$$

$$v = \sqrt{2gh - 4gR}$$

QUIZ 6.2

1. A 4.0-kg cart travels on a track at 6.0 m/s. A person pushes the cart along the direction of motion with a force of 8.0 N over a distance of 7.0 m. How fast does the cart travel now?
2. A 10-kg ball travels with a speed of 3.0 m/s. A person catches it and throws it back at a speed of 1.0 m/s. How much work did the person do on the ball?
3. A penguin slides down a frictionless hill that is 20 m high. The penguin starts from rest. What is its speed at the bottom of the hill?
4. In the previous problem, the penguin continues sliding and slows to 10 m/s. It then goes up a frictionless hill. What height will the penguin reach?

POTENTIAL ENERGY OF A SPRING

A spring has a natural length and can be stretched or compressed (Figure 6.5). The potential energy for a spring is

$$U = \tfrac{1}{2}kx^2$$

where k is the *spring constant*, and x is amount the spring has been compressed or stretched.

When $x = 0$, the spring is relaxed (at equilibrium) and $U = 0$.

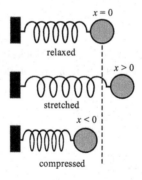

FIGURE 6.5   Spring and mass.

1. A spring has a spring constant of 320 N/m. Find the potential energy when it is stretched 0.23 m.

$U = \tfrac{1}{2}(320)(0.23)^2 = $ **8.5 J**

2. A 0.050-kg mass is attached to a spring ($k = 40$ N/m). The spring is compressed by 10 cm and then released. Find the speed of the mass at equilibrium ($x = 0$).

$E_i = \frac{1}{2}(40)(-0.10)^2 = 0.20 \text{ J}$

$E_f = \frac{1}{2}mv^2 = 0.025v^2$

$0.025v^2 = 0.20$

$v^2 = 8, v = \textbf{2.8 m/s}$

FRICTION AND THERMAL ENERGY

If there is kinetic (sliding) friction, then mechanical energy will be converted to thermal energy, and the object will heat up. The increase in thermal energy is

$$\Delta E_{th} = f_k \Delta x$$

where f_k is the magnitude of the kinetic friction, and Δx is the sliding distance. ΔE_{th} is added to the final energy E_f.

1. A 50-kg child slides down a 5.0-m high hill on a sled, with an initial speed 1.0 m/s. At the bottom of the hill, the child's speed is 8.0 m/s. Find the increase in thermal energy.

   $E_i = \frac{1}{2}mv^2 + mgh = \frac{1}{2}(50)(1) + (50)(9.8)(5) = 2475$

   $E_f = \frac{1}{2}mv^2 + \Delta E_{th} = \frac{1}{2}(50)(64) + \Delta E_{th} = 1600 + \Delta E_{th}$

   $2475 = 1600 + \Delta E_{th}$

   $\textbf{875 J} = \Delta E_{th}$

2. Starting from rest at a height $h = 3.0$, a block slides down a 20° inclined plane (Figure 6.6). The coefficient of kinetic friction is $\mu_k = 0.04$. Find the speed at the bottom.

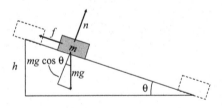

FIGURE 6.6   Block on an inclined plane with friction.

The sliding distance Δx is the hypotenuse.

$\sin \theta = h/\Delta x$

$\Delta x = h/\sin \theta = 3/\sin (20°) = 8.8 \text{ m}$

$\Delta E_{th} = (\mu_k \, mg \cos \theta)\Delta x$

$E_i = mgh$

$E_f = \frac{1}{2}mv^2 + \Delta E_{th}$

$\frac{1}{2}mv^2 + (\mu_k \, mg \cos \theta)\Delta x = mgh$

$v^2 = 2gh - 2(\mu_k g \cos \theta)\Delta x$

$v^2 = 2(9.8)(3) - (2)(0.04)(9.8)(0.94)(8.8) = 52.3$

$v = \textbf{7.2 m/s}$

QUIZ 6.3

1. A 5.0-kg mass is attached to spring, $k = 0.20$ N/m. The spring is stretched by 0.10 m. Find the potential energy.
2. In the previous problem, if the mass is released, what is the speed of the mass at equilibrium?
3. From an initial height of 3.0 m, a 40-kg child slides down a slide. The initial speed is 2.0 m/s. Because of friction, at the bottom of the slide the child's speed is only 1.0 m/s. What was the increase in thermal energy?
4. A block starts from rest and slides down a 30° inclined plane with $\mu_k = 0.20$. It slides a total distance $\Delta x = 5.0$ m. What is the final speed of the block?

CHAPTER SUMMARY

Work (J)	$W = F_{//}d$
Kinetic energy (J)	$K = \frac{1}{2}mv^2$
	$K_f = K_i + W$
Potential energy (J)	Gravity: $U = mgy$
	Spring: $U = \frac{1}{2}kx^2$
Energy conservation	$E = \frac{1}{2}mv^2 + U = \text{constant}$
Thermal energy due to friction	$\Delta E_{th} = f_k \Delta x$

END-OF-CHAPTER QUESTIONS

1. A proton of mass 1.67×10^{-27} kg travels at 6.0×10^4 m/s. An accelerator does 9.0×10^{-18} J of work on the proton. What is the final speed of the proton?
2. A box slides down an inclined plane. A person tries to prevent it from sliding by exerting a force of 5.0 N parallel to the plane. However, the box slides down the plane a distance of 3.0 m. How much work did the person do?

3. Starting from rest, a block slides down a 30° frictionless inclined plane. It slides a total distance of 4.0 m. What is its speed at the bottom?

4. After finals, a student opens a window of her dorm room, 10 m above the ground. She throws a water balloon upward with an initial speed 5.0 m/s.

    a. What is the speed of the balloon as it falls past the window?

    b. What is the speed of the balloon just before it hits the ground?

5. A 1200-kg car goes down a 50-m high hill at a constant speed. What was the increase in thermal energy?

ADDITIONAL PROBLEMS

1. A wrestler grabs a rhino's horn and pushes up with 14 N of force at an angle 44° above horizontal. The angry rhino pushes the wrestler back 3.0 m. How much work did the wrestler do?

2. Starting from rest, a block slides down a 60° inclined plane. It slides a total distance of 4.0 m. The coefficient of kinetic friction is 0.30. What is the speed of the block at the bottom?

3. *Power* (*P*) is the rate at which work is done. It is measured in units of J/s, also called Watts (W). The power output of a force is $P = F_{//}v$, where v is the object's speed. Suppose a boat is powered by an engine that provides a force of 140 N. The boat travels at 10 m/s. What is the power output of the engine?

4. Consider a 5.0-kg block on a 60° inclined plane with a coefficient of kinetic friction 0.30. A physics professor pushes the block up the plane at a constant speed of 3.0 m/s. What is the power output of the professor?

Rotational Motion

7

ANGULAR VELOCITY

An object rotates about an *axis* (Figure 7.1). Some objects, like the Earth, rotate slowly. Others, like a spinning top, rotate quickly. *Angular velocity* (ω) is measured in radians per second (rad/s), where a complete rotation is 2π radians. For a rigid object, all parts of the object have the same ω.

A point on the object has a tangential velocity, *v*. It is related to ω by

$$v = \omega R$$

where *R* is the distance from the point to the axis.

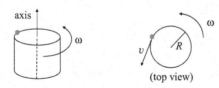

FIGURE 7.1 Object rotating about an axis.

1. A wheel with a 0.50-m radius rotates about a fixed axis at 120 rotations per minute (rpm). What is the angular velocity?

   Convert rpm to rad/s:

   $$\omega = \frac{120\,\text{rotations}}{\text{min}}\,\frac{1\,\text{min}}{60\,\text{s}}\,\frac{2\pi\,\text{rad}}{\text{rotation}} = 4\pi\,\text{rad/s}$$

2. In the previous problem, a piece of gum is stuck to the wheel. What is its tangential velocity?

$$v = \omega R = (4\pi)(0.5) = \textbf{6.28 m/s}$$

ROTATIONAL KINETIC ENERGY

The kinetic energy due to rotation is given by

$$K = \tfrac{1}{2} I \omega^2$$

where I is the *moment of inertia*, which has units kg m². Its value depends on the shape and density distribution of the object. A uniform object has a constant density throughout. In the next section, we will see how to calculate I.

1. A uniform disk has a moment of inertia given by $I = 1/2\, MR^2$. Consider a disk with $M = 2.0$ kg and $R = 0.40$ m. It spins at 19.1 rpm about its fixed central axis. Find the kinetic energy.

$$I = 1/2\, MR^2 = (0.5)(2)(0.4^2) = 0.16 \text{ kg m}^2$$

$$\omega = \frac{19.1 \text{ rotations}}{\text{min}} \frac{1 \min}{60 \text{ s}} \frac{2\pi \text{ rad}}{\text{rotation}} = 2.0 \text{ rad/s}$$

$$K = \tfrac{1}{2} I \omega^2 = (0.5)(0.16)(2^2) = \textbf{0.32 J}$$

2. A uniform sphere has a moment of inertia given by $I = 2/5\, MR^2$. A neutron star can be modeled as an extremely dense uniform sphere. Consider a neutron star of mass 1.0×10^{30} kg and radius 10 km. It spins at 5.0 rotations per second. What is its kinetic energy?

$$I = 2/5\, MR^2 = (0.4)(10^{30})(10{,}000^2) = 4.0 \times 10^{37} \text{ kg m}^2$$

$$\omega = \frac{5 \text{ rotations}}{\text{s}} \frac{2\pi \text{ rad}}{\text{rotation}} = 31.4 \text{ rad/s}$$

$$K = \tfrac{1}{2} I \omega^2 = (0.5)(4 \times 10^{37})(31.4^2) = \textbf{2.0} \times \textbf{10}^{\textbf{40}} \textbf{ J}$$

QUIZ 7.1

1. The Earth rotates once every 24 h. What is its angular velocity?
2. The radius of the Earth is approximately 6400 km. Find the tangential velocity of a person on the equator. (Ignore the motion of the Earth around the Sun.)
3. The Earth can be modeled as a uniform sphere ($I = 2/5 \, MR^2$) with a mass 6.0×10^{24} kg. Find its rotational kinetic energy.
4. Consider an optical disc ($I = 1/2 \, MR^2$) with a mass of 16 g and a diameter of 120 mm. It spins at 240 rpm. Find its kinetic energy.

MOMENT OF INERTIA

To find an object's moment of inertia, we divide the object into little masses: m_1, m_2, etc. (Figure 7.2). Each mass is some perpendicular distance (r_1, r_2, etc.) from the axis. The moment of inertia is defined as

$$I = m_1 r_1^2 + m_2 r_2^2 + \ldots$$

FIGURE 7.2   A piece m_1 of an object is a perpendicular distance r_1 from the axis.

1. Derive the moment of inertia for a hoop of mass, M, and radius, R (Figure 7.3).

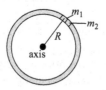

FIGURE 7.3   A hoop where the rotational axis points out of the page.

$$I = m_1 r_1^2 + m_2 r_2^2 + \ldots$$

$$= m_1 R^2 + m_2 R^2 + \ldots \text{ (All the little masses have the same } R)$$

$$= (m_1 + m_{2\,+} \,...)R^2$$

$I = \mathbf{MR^2}$ (The sum of the little masses is M)

2. Two small masses (0.10 kg) are connected by a massless rod of length 0.50 m (Figure 7.4). Calculate the moment of inertia if the rotational axis is in the center of the rod or if the axis goes through one of the masses.

FIGURE 7.4   Two masses connected by a rod. Two different rotational axes are shown.

Axis in the center:

$I = (0.1)(0.25^2) + (0.1)(0.25^2) = \mathbf{0.0125\ kg\ m^2}$

Axis through one of the masses:

$I = (0.1)(0^2) + (0.1)(0.5^2) = \mathbf{0.025\ kg\ m^2}$

TORQUE

Torque is a kind of twisting action that can cause objects to rotate. When you open a door, you exert a torque on the door that makes it rotate about its hinges. The rotational axis is called the *pivot* (for a door, it is the hinges). The magnitude of torque (τ) is

$$\tau = rF \sin \theta$$

where r is the distance from the pivot to the point where the force, F, is applied, and θ is the angle between $\mathbf{r}$ and $\mathbf{F}$ (Figure 7.5).

Torque has units of N·m. By convention, τ is positive if it pushes the object in a counterclockwise direction, and negative if it pushes in a clockwise direction.

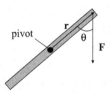

FIGURE 7.5   Quantities for calculating torque. $\mathbf{r}$ points from the pivot to the point where $\mathbf{F}$ is applied.

1. Find the torque when a person exerts 10 N on the edge of a door (Figure 7.6).

(top view)

F

1.0 m

FIGURE 7.6   Force exerted on a door (top view).

$$\tau = rF \sin 90° = rF$$

$$\tau = (1)(10) = \textbf{10 N m}$$

2. Find the torque when a person exerts 10 N on the center of the door (Figure 7.7).

$$\tau = rF = (0.5)(10) = \textbf{5.0 N m}$$

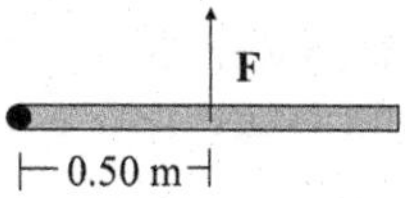

F

0.50 m

FIGURE 7.7   Force exerted on a door.

QUIZ 7.2

1. A hoop has a diameter of 1.0 m and mass of 0.60 kg. It rotates about its center. What is the moment of inertia?
2. Two small masses, 1.0 kg and 3.0 kg, are connected by a massless rod of length 4.0 m (Figure 7.8). The rotational axis is 1.0 m from the heavier mass. Find the moment of inertia.

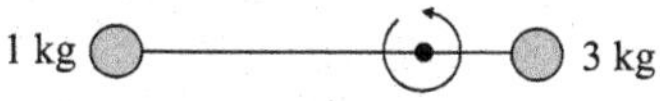

1 kg ⚪————— ⊙⚪ 3 kg

FIGURE 7.8   Two masses connected by a massless rod.

3. A person pushes the edge of a door with a force of 8.0 N and a 30° angle (Figure 7.9). What is the torque?

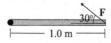

30° F

1.0 m

FIGURE 7.9   Force exerted on a door.

4. A driver wants to make a left turn. The steering wheel has a 0.20-m radius. Her left hand exerts a 3.0 N force down and her right hand exerts a 3.0 N force up (Figure 7.10). Find the torque.

FIGURE 7.10   Forces exerted on a steering wheel.

TURNING A WRENCH

Torque is important if you are using a wrench to loosen or tighten a bolt. In these problems, the bolt is the pivot.

1. A worker exerts 7.0 N of downward force on a wrench (Figure 7.11). The wrench is 0.20-m long and is oriented at 25° to horizontal. Calculate the torque.

FIGURE 7.11   Force on a wrench. The geometry is detailed on the right.

$r = 0.2$ m

$F = 7$ N

$\theta = 90° - 25° = 65°$

$\tau = (0.2)(7) \sin (65°) = \mathbf{1.3\ N\ m}$

The sign is *positive* because the torque makes the wrench rotate counterclockwise.

2. A worker exerts 8.0 N of force 30° with respect to horizontal (Figure 7.12). The wrench is vertical and has a length of 0.10 m. Find the torque.

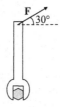

FIGURE 7.12   Force on a wrench.

$\theta = 30° + 90° = 120°$

$\tau = -(0.1)(8) \sin (120°) = \mathbf{-0.69\ N\ m}$

The sign is *negative* because the torque makes the wrench rotate clockwise.

ANGULAR ACCELERATION

Recall that acceleration (*a*) is change in velocity versus time and is measured in m/s² (Chapter 1). *Angular acceleration* (α) is change in angular velocity versus time and is measured in rad/s²:

$$\alpha = \Delta\omega/\Delta t$$

Torque causes angular acceleration:

$$\tau = I\alpha$$

This is the rotational analog of $F = ma$.

1. A merry-go-round ($I = 20$ kg m²) spins at 0.10 rad/s. A torque of 10 N m is applied, making it spin faster. How long must the torque be applied to make it spin at 2.1 rad/s?

$\tau = I\alpha$

$10 = (20)\ \alpha$

$\alpha = 0.5$ rad/s^2

$\alpha = \Delta\omega/\Delta t$

$0.5 = 2.0/\Delta t$ $(\Delta\omega = 2.1 - 0.1 = 2.0$ rad/s$)$

$\Delta t = 2/0.5 = $ **4.0 s**

2. String is wrapped around the inner diameter ($r = 0.030$ m) of a spool (Figure 7.13). The spool has a moment of inertia $I = 1/2\ MR^2$, where $M = 0.50$ kg and $R = 0.10$ m. It is free to rotate about a fixed axis. A cat pulls the string with 2.0 N of force. Find the angular acceleration.

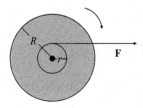

FIGURE 7.13 Force exerted on a string that is wrapped around a cylinder of radius, r.

$I = 1/2\ MR^2 = (0.5)(0.5)(0.1^2) = 0.0025$ kg m^2

$\tau = rF = -(0.03)(2) = -0.06$ N m ($-$ because it pulls the spool in a clockwise direction)

$\tau = I\alpha$

$-0.06 = (0.0025)\ \alpha$

$\alpha = $ **−24 rad/s^2**

QUIZ 7.3

1. A worker exerts a downward force of 10 N on a wrench that is 30° to horizontal (Figure 7.14). The length of the wrench is 0.15 m. Find the torque.

FIGURE 7.14 Force on a wrench.

2. An aluminum rod hangs vertically (Figure 7.15). A force of 50 N is applied, 35° to horizontal and 1.5 m from the pivot. Calculate the torque.

FIGURE 7.15   Force on a hanging rod.

3. Bobby has a globe ($I = 2/5\,MR^2$, $M = 1.0$ kg, $R = 0.20$ m) that is spinning at 4.0 rad/s. He puts his finger on the globe, and it stops spinning in 3.0 s. What was the magnitude of the torque that Bobby applied?

4. Rope is wound around a cylinder ($I = 1/2\,MR^2$, $M = 48$ kg, $R = 0.50$ m) that is free to rotate about a fixed axis (Figure 7.16). A sailor pulls the rope with a force of 60 N. Find the angular acceleration.

FIGURE 7.16   Force exerted on a string that is wrapped around a cylinder of radius, R.

CHAPTER SUMMARY

Angular velocity (ω, rad/s)	$v = \omega R$
Tangential velocity (v, m/s)	
Rotational kinetic energy (J)	$K = \tfrac{1}{2}I\omega^2$
Moment of inertia (kg·m²)	$I = m_1 r_1^2 + m_2 r_2^2 + \ldots$
Torque (N·m)	$\tau = rF\sin\theta$
	Counterclockwise +, clockwise –
Angular acceleration (α, rad/s²)	$\alpha = \Delta\omega/\Delta t$
	$\tau = I\alpha$

END-OF-CHAPTER QUESTIONS

1. A merry-go-round with a diameter of 4.0 m rotates at 15 rpm.
   a. What is its angular velocity?
   b. A child is on the edge. What is the child's velocity?

2. An airplane is preparing for takeoff. Its propeller ($I = 1/3\ MR^2$) rotates at 1200 rpm. The mass of the propeller is 100 kg and its radius is 1.0 m. Find the rotational kinetic energy.
3. Two masses, 3.0 kg and 5.0 kg, are attached by a 4.0-m massless rod. They rotate at 4.0 rad/s about a fixed central axis. Calculate the kinetic energy.
4. At a factory, a very long (3.0 m) wrench is used to tighten a large bolt (Figure 7.17). The wrench is at 40° to horizontal. A worker (100 kg) tries to turn the wrench by hanging from the end. What is the torque?

FIGURE 7.17   Person hanging from a huge wrench.

5. Two 2.0-kg masses are connected by a massless rod with a length of 1.5 m (Figure 7.18). At the rod's center is massless cylinder ($r = 0.40$ m) with a central axis. String is wound around the cylinder. A person pulls on the string with a force of 5.0 N. Find the angular acceleration.

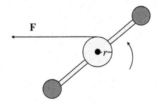

FIGURE 7.18   Two masses connected by a massless rod and cylinder.

ADDITIONAL PROBLEMS

1. A massless rod of length $2a$ ($a = 2.0$ m) has identical masses ($m = 3.0$ kg) attached. The masses are a and $2a$ from the rotation axis (Figure 7.19). The rod is tilted 60° above horizontal.
   a. What is the torque about the rotation axis?
   b. After some time, the rod and masses rotate at 6.0 rpm. What is the kinetic energy?

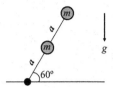

FIGURE 7.19   Two masses connected by a massless rod.

2. A ball with a mass of 6.0 kg and a radius of 0.30 m spins at 8.0 rad/s. A person catches it. The sliding friction from the person's hands makes the ball stop spinning in 0.10 s. What was the average torque (magnitude) exerted by the hands? Model the ball as a uniform sphere, $I = 2/5\ MR^2$.

3. Rope is wound around a cylinder ($I = 1/2\ MR^2$) with a mass of 64 kg and a radius of 0.50 m. The cylinder is initially at rest. A sailor pulls the rope with a force of 8.0 N over a distance 2.0 m.
   a. How much work was done by the sailor?
   b. What is the angular velocity of the cylinder?

4. Two identical twins (30 kg each) are on a merry-go-round. One twin is 1.0 m from the center, and the second twin is 2.0 m from the center. The merry-go-round is a disk with a radius of 3.0 m and has a mass of 100 kg, $I = 1/2\ MR^2$. The merry-go-round spins at 15 rpm.
   a. Find the moment of inertia for the merry-go-round with the children on it.
   b. Calculate the kinetic energy.

Momentum

MOMENTUM AND IMPULSE

Momentum (**p**) is defined as mass times velocity:

$$\mathbf{p} = m\mathbf{v}$$

Like velocity, **p** is a vector that points in a certain direction.

Force is required to change the momentum of an object. Let $\mathbf{F}_{av}$ be the average force exerted on an object during a time interval, Δt. The change in momentum ($\Delta \mathbf{p}$), called the *impulse*, is given by

$$\Delta \mathbf{p} = \mathbf{F}_{av}\Delta t$$

1. During a safety test, a 500-kg car travels at 20 m/s in the x direction (Figure 8.1). It crashes into a wall and comes to rest. What was the impulse?

FIGURE 8.1 Car crashing into a wall.

We consider the component of **p** in the x direction.

Before: $p_i = (500)(20) = 10,000$ kg·m/s

After: $p_f = (500)(0) = 0$

$\Delta p = p_f - p_i = \mathbf{-10,000}$ **kg·m/s**

2. In the previous question, the car took 0.20 s to go from 20 m/s to 0. What was the average force exerted by the wall during that time?

$$\Delta p = F_{av}\Delta t$$

$$-10{,}000 = F_{av}(0.2)$$

$$\mathbf{-50{,}000\ N} = F_{av}$$

INELASTIC COLLISIONS

When objects collide, the momentum of the system is conserved, so long as there is no external force (e.g., a physics professor holding an object to keep it from moving). The sum of the objects' momenta just before the collision equals the sum of momenta just after the collision. A collision is *inelastic* if the objects stick together or there is friction.

1. A 700-kg car travels at 14 m/s and collides with an identical car that is parked (Figure 8.2). The cars stick together. What is the speed of the cars immediately after the collision?

before   after

FIGURE 8.2   Collision between two identical cars.

$$p_i = mv_0$$

$$p_f = (2m)v$$

Momentum conservation: $mv_0 = (2m)v$

$$v_0/2 = v$$

$$\mathbf{7.0\ m/s} = v$$

2. In the previous question, what was the increase in thermal energy?

$$E_i = \tfrac{1}{2}mv_0^2 = (0.5)(700)(14^2) = 68{,}600$$

$$E_f = \tfrac{1}{2}(2m)v^2 + \Delta E_{th} = 34{,}300 + \Delta E_{th}$$

$$68{,}600 = 34{,}300 + \Delta E_{th}$$

$$\mathbf{34{,}300\ J} = \Delta E_{th}$$

QUIZ 8.1

1. A 2000-kg jet travels at 200 m/s in the x direction. An afterburner pushes the jet for 5.0 s. The jet reaches a speed of 250 m/s. What was the change in momentum?
2. In the previous problem, what was the average force exerted by the afterburner?
3. A person throws a 0.50-kg lump of clay at a speed of 10 m/s. The clay sticks to a 9.5-kg stationary cart. What is the speed of the clay plus cart immediately after the collision?
4. In the previous problem, what was the increase in thermal energy?

ELASTIC COLLISIONS

If a collision is *elastic*, then no thermal energy is generated. Objects such as billiard balls or atoms, which "bounce" rather than stick, undergo elastic collisions.

1. A 1.0-kg mass travels 2.0 m/s in the x direction (Figure 8.3). It collides with a 3.0-kg mass that is initially at rest. After the collision, the 1.0-kg mass travels 1.0 m/s in the $-x$ direction. Find the velocity of the 3.0-kg mass.

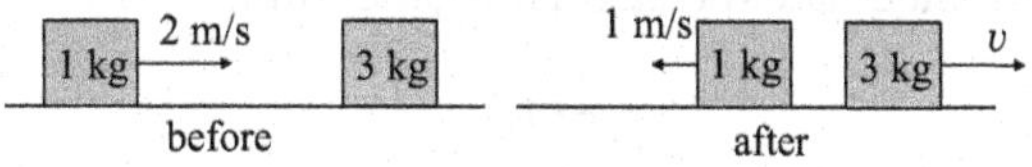

FIGURE 8.3   Collision between two masses.

$$p_i = (1)(2) + (3)(0) = 2$$

$$p_f = (1)(-1) + 3v = -1 + 3v$$

Momentum conservation: $2 = -1 + 3v$

$$v = \textbf{1.0 m/s}$$

2. In the previous problem, how much thermal energy was generated?

$$E_i = \tfrac{1}{2}(1)(2^2) = 2$$

$$E_f = \tfrac{1}{2}(1)(-1)^2 + \tfrac{1}{2}(3)(1)^2 + \Delta E_{th} = 2 + \Delta E_{th}$$

$$2 = 2 + \Delta E_{th}$$

$$\Delta E_{th} = \textbf{0}$$

Because no thermal energy was generated, the collision was elastic.

ANGULAR MOMENTUM

Suppose we have a disk rotating about a fixed central axis. Because the axis is fixed, the object has 0 momentum, **p**. However, it does have *angular momentum* (*L*), defined as

$$L = I\omega$$

This is the rotational analog of $\mathbf{p} = m\mathbf{v}$. By convention, L is positive if the rotation is counterclockwise, and negative if it is clockwise.

1. A uniform disk ($I = 1/2\,MR^2$) with a mass of 0.10 kg and a radius of 0.10 m spins at 120 rpm in a clockwise direction. Find the angular momentum.

   $I = 1/2\,MR^2 = (0.5)(0.1)(0.1^2) = 5 \times 10^{-4}\ \text{kg·m}^2$

   $$\omega = \frac{120\ \text{rotations}}{\text{min}}\ \frac{1\,\text{min}}{60\ \text{s}}\ \frac{2\pi\ \text{rad}}{\text{rotation}} = 12.6\ \text{rad/s}$$

   $L = -(5 \times 10^{-4})(12.6) = \mathbf{-0.0063\ kg·m^2/s}$

   (– because rotation is clockwise)

2. A small mass, m, travels in a circle of radius, R, with tangential velocity, v (Figure 8.4). Calculate the angular momentum.

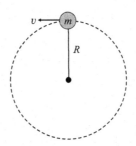

FIGURE 8.4   Mass undergoing uniform circular motion.

   The moment of inertia is $I = mR^2$.

   The tangential velocity is $v = \omega R$, or $\omega = v/R$.

   $L = (mR^2)(v/R) = \mathbf{mvR}$

QUIZ 8.2

1. A 1.0-kg mass travels 3.0 m/s in the x direction. A 5.0-kg mass travels 3.0 m/s in the $-x$ direction. They collide. After the collision, the 1.0-kg mass travels 7.0 m/s in the $-x$ direction. Find the velocity of the 5.0-kg mass.

2. In the previous problem, calculate the increase in thermal energy as a result of the collision.
3. Consider Earth to be a uniform sphere ($I = 2/5\ MR^2$) with a mass of 6.0×10^{24} kg and a radius of 6400 km. It rotates once every 24 h. What is the angular momentum (as viewed from a point above the North Pole)?
4. A race car with a mass of 1000 kg travels at 30 m/s around a circular track with a radius of 250 m. The car travels in a clockwise direction. Find the angular momentum.

CONSERVATION OF ANGULAR MOMENTUM

If there is no external torque on a rotating object, then angular momentum is conserved. For example, an ice skater brings in her arms, reducing her moment of inertia (I). To conserve angular momentum ($L = I\omega$), her angular velocity (ω) must increase.

1. A massless rod of length 1.0 m has two masses attached to it (Figure 8.5). It rotates at 3.0 rad/s. The rod expands to a length of 2.0 m. Find the angular velocity.

before   after

FIGURE 8.5   Rotation of a rod with two masses attached.

$I = mR^2 + mR^2 = 2mR^2$

$L_i = (2m \cdot 0.5^2)(3) = 1.5m$

$L_f = (2m \cdot 1.0^2)\omega = 2m\omega$

Angular momentum conservation: $1.5m = 2m\omega$

$1.5 = 2\omega$

0.75 rad/s $= \omega$

2. A star ($I = 2/5\ MR^2$) with a radius of 1.0×10^6 km rotates at 1.0×10^{-5} rad/s. It collapses to a neutron star, which has a radius of 10 km. Assuming no loss of mass, find the angular velocity.

$L_i = 2/5\ M(10^9)^2(1 \times 10^{-5}) = 4 \times 10^{12}\ M$

$L_f = 2/5\ M(10^4)^2\omega = 4 \times 10^7\ M\ \omega$

Angular momentum conservation: $4 \times 10^{12}\ M = 4 \times 10^7\ M\ \omega$

1.0×10^5 rad/s $= \omega$

SPINNING DISKS

These problems involve the angular momentum conservation of a spinning disk plus a small mass. The important point is to calculate the angular momentum resulting from the disk *and* the mass.

1. A uniform disk ($I = 1/2\ MR^2$) with a mass of 4.0 kg and a radius of 0.50 m spins at $\omega_0 = 3.0$ rad/s (Figure 8.6). A piece of clay ($m = 0.25$ kg) is dropped onto the edge of the disk. Find the angular velocity of the disk plus clay.

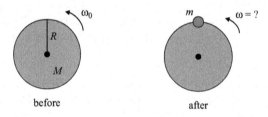

before   after

FIGURE 8.6   Piece of clay dropped onto a spinning disk.

Before: $I = 1/2\ MR^2 = 0.5$

$L_i = I\omega_0 = (0.5)(3) = 1.5$

After: $I = 1/2\ MR^2 + mR^2 = 0.5 + 0.0625 = 0.5625$

$L_f = I\omega = 0.5625\omega$

Angular momentum conservation: $1.5 = 0.5625\omega$

$\omega = \mathbf{2.7\ rad/s}$

2. A person ($m = 50$ kg) is on a merry-go-round ($I = 1/2\ MR^2$, $M = 100$ kg, $R = 3.0$ m) initially at rest. The person runs at a tangential velocity, $v = 3.0$ m/s, relative to the ground (Figure 8.7). Find the angular velocity of the merry-go-round.

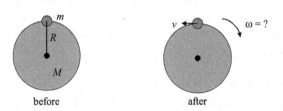

before   after

FIGURE 8.7   Person on a merry-go-round.

Before: $L_i = 0$

After: $L(\text{person}) = mvR = 450$

$L(\text{merry-go-round}) = -1/2\, MR^2 \omega = -450\omega$

Total $L_f = 450 - 450\omega$

Angular momentum conservation: $0 = 450 - 450\omega$

$\omega = \mathbf{1.0\ rad/s}$

QUIZ 8.3

1. A person holds two 50-kg masses at arm's length (0.80 m from the center). The person sits on a chair that rotates at 1.5 rad/s. The person brings his arms in so the masses are only 0.20 m from the center. What is the angular velocity? Ignore the moment of inertia of the person and the chair.
2. In the previous problem, suppose the chair plus person has a moment of inertia 36 kg·m². (This must be added to the moment of inertia of the two 50-kg masses.) Find the angular velocity.
3. A 75-kg person is at the center of a 150-kg merry-go-round ($I = 1/2$ MR^2, $R = 2.0$ m), which spins at 1.5 rad/s. The person walks to the edge of the merry-go-round. Now what is the angular velocity?
4. A 40-kg child is on a 150-kg merry-go-round ($I = 1/2\, MR^2$, $R = 2.0$ m), initially at rest. She runs in a circle of radius 1.0 m with a speed 0.5 m/s relative to the ground. Calculate the angular velocity of the merry-go-round.

CHAPTER SUMMARY

Momentum	$p = mv$
Impulse	$\Delta p = F_{av} \Delta t$
Angular momentum	$L = I\omega$
	$L = mvR$
	Counterclockwise +, clockwise −

END-OF-CHAPTER QUESTIONS

1. A 0.50-kg basketball hits the floor with a speed 7.0 m/s. It is in contact with the floor for 0.020 s. It bounces back in the opposite direction with a speed 5.0 m/s. What was the average force exerted by the floor?
2. A 2.0-kg cart and a 1.0-kg cart approach each other, each with a speed 3.0 m/s. They collide and stick together.
   a. Find the speed of the carts after the collision.
   b. How much thermal energy was generated?

3. Starting from rest, a car rolls down a 20-m high frictionless hill. At the bottom, it collides with an identical parked car. They stick together. What is their speed?

4. Two small, identical masses are in a 1.0-m long tube of negligible mass (Figure 8.8). The masses are connected to the central axis by two 0.20-m long strings. The tube rotates at 8.0 rad/s. One string breaks, causing the mass to slide to the end of the tube. Calculate the angular velocity.

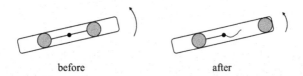

before   after

FIGURE 8.8   Masses in a rotating tube.

5. Initially at rest, a ballerina ($I = 20$ kg·m²) is on a raft ($I = 100$ kg·m²) on a lake. Suddenly, she spins at 20 rad/s clockwise. Find the angular velocity of the raft. Ignore friction.

ADDITIONAL PROBLEMS

1. A 45-kg ball is suspended by a rope. A 30-kg child runs 2.0 m/s and grabs on to the ball. How high will the child and ball swing?

2. A 1.0-kg object moves at 9.0 m/s toward a 3.0-kg object. They collide and are motionless.
   a. What was the speed of the 3.0-kg object before the collision?
   b. What was the increase in thermal energy?

3. When Halley's Comet is furthest from the Sun, a distance of 35 AU, it travels 0.88 km/s. (One AU is the distance from Earth to the Sun.) Its closest distance to the Sun is 0.59 AU. Use angular momentum conservation to find its speed at 0.59 AU.

4. Two uniform disks ($I = 1/2\ MR^2$) are spinning about the same central axis. They both have a radius of 0.10 m. The first disk is 2.0 kg and spins clockwise at 15 rpm. The second disk is 4.0 kg and spins counterclockwise at 40 rpm. They collide and stick together. What is their rotational speed (rpm)?

Oscillations and Waves

HOOKE'S LAW

Consider a mass attached to a spring. The spring is relaxed and not stretched or compressed. The mass is at its *equilibrium* position, denoted $x = 0$ (Figure 9.1). A person pulls on the mass so that its position is $x > 0$. The spring is now stretched.

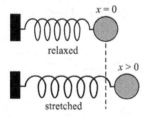

FIGURE 9.1 Mass attached to a spring.

According to Hooke's law, the spring exerts a force on the mass,

$$F = -kx$$

where k is the *spring constant*. The minus sign means if the mass has a positive x value, the spring will pull in the $-x$ direction. If the mass has a negative x value, the spring will push in the $+x$ direction.

1. A person pulls a spring with 10 N of force, causing it to stretch 0.10 m in the x direction. What is the spring constant?

   The spring exerts a force of 10 N in the $-x$ direction:

   $$F = -kx$$

$-10 = -k(0.1)$

100 N/m $= k$

2. A spring ($k = 50$ N/m) hangs vertically. A 2.0-kg mass is attached. How much does the spring stretch?

Magnitude of gravity force: $mg = (2)(9.8) = 19.6$ N

Magnitude of spring force: $kx = (50)x$

These forces must balance: $19.6 = 50x$

0.39 m $= x$

OSCILLATIONS

Suppose a person pulls the mass to $x = A$ (Figure 9.2). Then, at $t = 0$, the person lets go. It will *oscillate* back and forth. In one cycle, the mass goes from A to $-A$ and back to A.

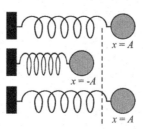

FIGURE 9.2   Oscillating mass and spring. The dashed line is $x = 0$.

A is the *amplitude* in meters (m). ω is the *angular frequency* (rad/s) and tells how rapidly the mass oscillates. For a spring and mass,

$$\omega = \sqrt{k/m}$$

The *frequency* (f) is the number of times the mass goes back and forth each second:

$$f = \frac{\omega}{2\pi}$$

Frequency has units of s^{-1}, also called Hertz (Hz), or cycles per second. The *period* (*T*) is the time it takes (s) to complete one cycle:

$$T = 1/f$$

1. A 4.0-kg mass is attached to a spring, $k = 16$ N/m. It is displaced 5.0 cm from equilibrium and released. Find the amplitude, frequency, and period of its oscillation.

   $A = 5.0$ cm

   $\omega = \sqrt{16/4} = 2$ rad/s

   $f = \omega/(2\pi) = 0.32$ s^{-1}

   $T = 1/f = 3.1$ s

2. A 0.10-kg mass hangs vertically from a spring with a spring constant 10 N/m. A person pulls it down 2.0 cm and lets go. How long does it take to reach the maximum height?

   $\omega = \sqrt{10/0.1} = 10$ rad/s

   $f = \omega/(2\pi) = 1.6$ s^{-1}

   $T = 1/f = 0.628$ s

   The mass goes from $y = -2$ cm to $y = 2$ cm. The time it takes is half the period.

   $T/2 = 0.31$ s

QUIZ 9.1

1. A mass is attached to a spring, $k = 17$ N/m. Someone pushes the mass 0.30 m in the $-x$ direction, compressing the spring. Find the force exerted by the spring on the mass.
2. A 1.0-kg mass hangs vertically from a spring. A 2.0-kg mass is attached to the 1.0-kg mass. This causes the spring to stretch by an additional 0.10 m. Find the spring constant.
3. A 10-g mass is attached to a spring with a spring constant 10 N/m. It is displaced 7.0 cm from equilibrium and released. Find the amplitude, frequency, and period of the oscillation.

4. A 2.0-kg mass hangs vertically from a spring, $k = 20$ N/m. The professor raises the mass 5.0 cm and releases it. How long does it take to
reach the equilibrium position ($y = 0$)?

MATHEMATICAL DESCRIPTION

Suppose the mass is stretched to $x = A$ and released at $t = 0$. Its position (x) for
$t > 0$ is given by a cosine function (Figure 9.3). Its velocity (v) and acceleration (a)
are described similarly. Note that ωt is in radians and *not* degrees.

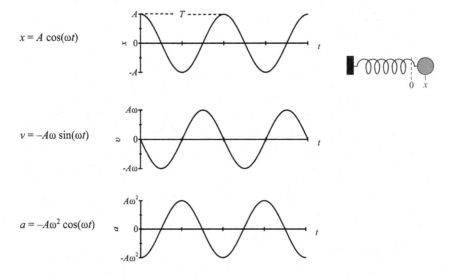

$x = A \cos(\omega t)$

$v = -A\omega \sin(\omega t)$

$a = -A\omega^2 \cos(\omega t)$

FIGURE 9.3   Plots of position, velocity, and acceleration for a mass attached to
a spring.

1. A 0.10-kg mass is attached to a spring, $k = 10$ N/m. It is displaced
0.20 m from equilibrium and released. What is its maximum speed
and where does it occur?

$v = -A\omega \sin(\omega t)$

Because sin is between −1 and 1, the maximum speed is $\mathbf{v_{max} = A\omega}$

$\omega = \sqrt{10/0.1} = 10$ rad/s

$v_{max} = (0.2)(10) = \mathbf{2.0 \ m/s}$

From the graphs, the maximum speed occurs when $\mathbf{x = 0}$ (equilibrium).

2. In the previous problem, what is the maximum acceleration magnitude, and where does it occur?

$a = -A\omega^2 \cos(\omega t)$

Because cos is between −1 and 1, the maximum acceleration is $A\omega^2 =$ (0.2)(100) = **20 m/s²**
The maximum acceleration magnitude occurs when $x = -A$ **or** A (when the spring is maximally compressed or stretched).

x = **−0.20 m or 0.20 m**

TRAVELING WAVES

Consider a long string under tension. At the left end, someone wiggles the string up and down. This causes a *transverse wave* to travel toward the right. It is called transverse because the atoms of the string oscillate vertically, and the wave travels horizontally. Figure 9.4 shows a snapshot of the wave. A movie would show the wave pattern moving toward the right.

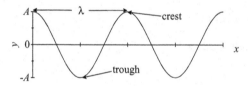

FIGURE 9.4   Plot of a wave at a particular time, *t*.

The maximum of a wave is a *crest*, and the minimum is a *trough*. The height of the crest is the *amplitude* (*A*). The distance from crest to crest is the *wavelength* (λ). The wavelength is the distance over which the wave pattern repeats.
A cosine wave is

$$y = A\cos\left(\frac{2\pi}{\lambda}x \pm \omega t\right)$$

The − is for waves that travel right, and + is for waves that travel left. The quantity in the parentheses is in radians. The *frequency* gives the number of crests that go by each second:

$$f = \frac{\omega}{2\pi}$$

The *wave speed* (*v*) is

$$v = \lambda f$$

1. A wave on a string travels 0.50 m in 0.10 s as shown in Figure 9.5. What are the amplitude, wavelength, and wave speed?

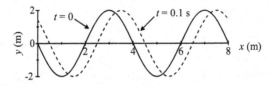

FIGURE 9.5   Plot of a wave for two values of t.

Amplitude = **2.0 m**

Wavelength = **4.0 m**

Wave speed = 0.5 m/0.1 s = **5.0 m/s**

2. A wave is given by $y = A \cos(6x - 2058t)$, where x is in meters, and t is in seconds. Find the wavelength, frequency, wave speed, and direction.

$$\frac{2\pi}{\lambda} = 6$$

$\lambda = 2\pi/6 = $ **1.05 m**

$f = 2058/(2\pi) = $ **328 s⁻¹**

$v = \lambda f = $ **340 m/s**

The − means the wave travels **right (+x direction)**

QUIZ 9.2

1. A 0.40-kg mass is attached to a spring, $k = 10$ N/m. It is displaced 5.0 cm from equilibrium and released. What is its speed at equilibrium ($x = 0$)?
2. In the previous problem, find the acceleration of the mass when the spring is maximally compressed.
3. A snapshot of a water wave is shown in Figure 9.6. Find the amplitude and wavelength.

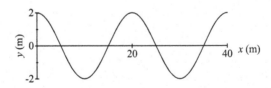

FIGURE 9.6   Plot of a water wave.

4. In the previous problem, suppose the wave travels toward the right at 4.0 m/s. Write an equation for the wave.

STANDING WAVES

Consider a stringed instrument like a violin or guitar. The string is clamped, or fixed, at two ends. A vibration of the string is a *standing wave*. The two ends are *nodes*, points where the string does not move. The sine wave equals zero at the nodes. An *antinode* is a point of maximum displacement.

A specific type of standing wave is a *mode*. The longest wavelength mode is the *fundamental* (Figure 9.7). For our string, it's half a sine wave. The wavelength is therefore $2L$. The figure shows a snapshot of the string (solid curve). At a later time, it is inverted (dashed curve). The string oscillates between these two profiles.

The fundamental is called the $m = 1$ mode. The $m = 2$ mode is a sine wave with wavelength L. It has a node in the center. The $m = 3$ mode has an additional node and has wavelength $2L/3$.

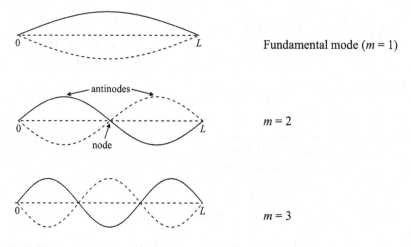

Fundamental mode ($m = 1$)

$m = 2$

$m = 3$

FIGURE 9.7   Standing waves.

1. An instrument has a 1.0-m long string that has a fundamental mode of 440 Hz. What is the wave speed (the speed of a traveling wave on this string)?

$\lambda = 2L = 2 \text{ m}$

$f = 440 \text{ Hz}$

$v = \lambda f = \mathbf{880 \ m/s}$

2. The wave speed on a string is 12 m/s. The string is clamped at two points, 3.0 m apart. Find the frequencies of the first three modes.

$\lambda = 2L, L$, and $2L/3$

$\quad = 6, 3$, and 2 m

$v = \lambda f$, or $f = v/\lambda$

$f = 12/6, 12/3, 12/2 = \textbf{2.0, 4.0, 6.0 Hz}$

INTERFERENCE

Consider two speakers on the x axis that emit sound waves of the same frequency (Figure 9.8). (The crest represents a region of high-density air, and the trough is low density). A listener is also on the x axis. The wave from each speaker travels a different distance, or path length.

<- 2λ -> constructive interference   <- $2\frac{1}{2}\lambda$ -> destructive interference

FIGURE 9.8   Sound waves travelling from two speakers.

If the path-length difference is a multiple of λ ($0, \lambda, 2\lambda$ ...), then we have *constructive interference*. The crests of one wave line up with the crests of the other wave. The troughs line up with the troughs. The addition (superposition) of the two waves is a big wave. The listener hears a loud sound.

Suppose the path-length difference is a multiple of λ plus ½ (½λ, 1½ λ, 2½ λ ...). In that case, we have *destructive interference*. The crests line up with the troughs. The two waves cancel each other out, so the sound is quiet.

1. Two speakers on the x axis emit the same 500-Hz tone. A person on the x axis thinks the sound is annoying. What should the minimum separation between the speakers be to make the sound quiet? The speed of sound in air is 343 m/s.

$v = \lambda f$, or $\lambda = v/f$

$\lambda = 343/500 = 0.686$ m

The speakers should be $\lambda/2$ apart for destructive interference.

$\lambda/2 = \textbf{0.34 m}$

2. Two speakers emit identical 686-Hz sound waves (Figure 9.9). Does the person hear a quiet or loud sound?

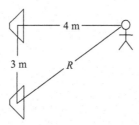

FIGURE 9.9   Person listening to sound from two speakers.

$\lambda = v/f = 343/686 = 0.5$ m

$R = \sqrt{3^2 + 4^2} = 5$ m

The path-length difference is 5 m – 4 m = 1 m. This is 2 λ.
The person is at a point of **constructive interference**, so the sound is **loud**.

QUIZ 9.3

1. A 2.0-m long string is clamped at both ends. Its fundamental mode has a frequency of 300 Hz. What is the wave speed?
2. An instrument has a 0.40-m long string. The wave speed is 200 m/s. Find the frequencies of the first three modes.
3. Two speakers and a listener are along a common axis. The speakers emit identical sound waves of frequency 654 Hz. Because of their size, the speakers must be at least 1.2 m apart. What is the minimum distance between the speakers such that the listener will experience constructive interference?
4. A person listens to two speakers that emit identical 257-Hz tones (Figure 9.10). What is the path-length difference? Is the person in a loud or quiet spot?

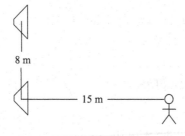

FIGURE 9.10   Person listening to sound from two speakers.

CHAPTER SUMMARY

Hooke's law	$F = -kx$
Angular frequency (rad/s), spring and mass	$\omega = \sqrt{k/m}$
Frequency (Hz or s^{-1})	$f = \dfrac{\omega}{2\pi}$
Period (s)	$T = 1/f$
Oscillation	$x = A \cos(\omega t)$
	$v = -A\omega \sin(\omega t)$
	$a = -A\omega^2 \cos(\omega t)$
Wave speed (m/s)	$v = \lambda f$
Traveling wave	$y = A\cos\left(\dfrac{2\pi}{\lambda}x \pm \omega t\right)$
Constructive interference	Path-length difference $= 0, \lambda, \ldots$
Destructive interference	Path-length difference $= \frac{1}{2}\lambda, 1\frac{1}{2}\lambda, \ldots$

END-OF-CHAPTER QUESTIONS

1. A spring is attached to the ceiling and hangs vertically. When a 1.0-kg mass is attached, the spring stretches by 0.50 m. The mass then undergoes oscillations. Find the frequency.

2. A 0.10-kg mass is attached to a spring. The spring constant is 10 N/m. The mass oscillates with an amplitude of 7.0 cm.
   a. What is the maximum speed of the mass, and where does this occur?
   b. What is the maximum potential energy, and where does it occur?

3. A wave on a string is given by $y = (7.0 \text{ cm}) \cos(4\pi x + 12\pi t)$, where x is in meters and t is in seconds. Find the amplitude, wavelength, frequency, wave speed, and direction.

4. A student determines that a traveling wave on a clamped 5.0-m long string has a speed of 20 m/s. What are the frequencies of the first three standing-wave modes?

5. Two speakers are 50 m and 70 m away from a listener. The speakers play identical 68.6-Hz tones. Is the listener in a quiet or loud spot? The speed of sound is 343 m/s.

ADDITIONAL PROBLEMS

1. A 1.0-kg mass is attached to a relaxed horizontal spring, $k = 4.0$ N/m. A 3.0-kg mass collides with the 1.0-kg mass and sticks to it. How long does it take for the spring to become fully compressed?

2. A simple pendulum is a small mass attached to a massless rod or string. Its angular frequency of oscillation is $\omega = \sqrt{g/L}$, where L is the length of the rod. A mass hangs from a 10-cm string and swings back and forth. Find the frequency and period of the oscillation.

3. A water wave on a pond has an amplitude 20 cm, speed 1.0 m/s, and wavelength 2.0 m. At $t = 0$, a duck sits in a trough of the wave ($y = -20$ cm). What is the duck's height y at $t = 0.20$ s?

4. Water waves with a wavelength of 0.10 m travel through two small holes that are 0.50 m apart (Figure 9.11). A bug floats on the water.

   a. What is the path-length difference for water waves traveling from the two holes to the bug?

   b. Will the bug's oscillations be large or small?

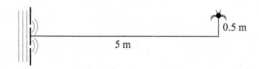

5 m

0.5 m

FIGURE 9.11   Bug on the water (top view).

Fluids

PRESSURE

Pressure (P) is defined as force per area,

$$P = F/A$$

P is measured in units of N/m², or Pascals (Pa). One atmosphere of pressure (1 atm) is 101.3 kPa = 1.013×10^5 Pa. Right now, gas molecules are colliding with your skin. These collisions produce 1 atm of pressure, which is quite a lot. Fortunately, your insides (e.g., lungs, etc.) are at an equal pressure, so you don't implode!

In a liquid such as water, the pressure increases with depth because of the weight of the liquid above that depth. If the liquid is not moving, the pressure is given by

$$P = P_0 + \rho g d$$

where P_0 is the pressure at the surface, ρ is the density of the liquid (kg/m³), and d is the depth (m). This equation does not apply to gases. The density of water is 1000 kg/m³ (i.e., a cubic meter has a mass of 1000 kg.)

1. What is the force exerted by the air on one side of a 2.0-m × 2.0-m × 2.0-m cube?

   $P = F/A$, or $F = PA$

   The air pressure is $P = 1.013 \times 10^5$ Pa.

   The area of one side is $A = 2 \times 2 = 4$ m²

   $F = (1.013 \times 10^5)(4) = \mathbf{4.0 \times 10^5}$ **N**

2. A scuba diver is at a depth of 20.0 m. What is the water pressure?

The pressure at the surface is 1 atm: $P_0 = 1.013 \times 10^5$ Pa.

$P = P_0 + \rho gd$

$P = 1.013 \times 10^5 + (1000)(9.8)(20) = \mathbf{2.97 \times 10^5}$ **Pa** (about 3 atm)

STATIC FLUIDS

A *fluid* is defined as a substance that does not resist deformation. We can place it in a container of any shape. Examples include air, water, and oil. A *static fluid* is a fluid that is not moving.

A connected liquid is one where the liquid molecules are free to move from one region to another. In a connected static liquid, if we draw a horizontal line, the pressure is the same everywhere along the line (Figure 10.1).

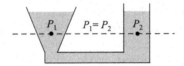

FIGURE 10.1   The pressure in a connected liquid is the same along a horizontal line.

1. A U-tube (Figure 10.2) contains liquid mercury ($\rho = 1.36 \times 10^4$ kg/m³). One side is a vacuum ($P = 0$), and the other side has a pressure 1 atm. Find the height difference, h.

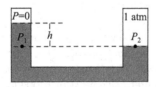

FIGURE 10.2   U-tube with liquid mercury.

$P_1 = 0 + \rho gh = (1.36 \times 10^4)(9.8)h$

$P_2 = 1$ atm $= 1.013 \times 10^5$ Pa

Set them equal: $(1.333 \times 10^5)h = 1.013 \times 10^5$

$h = \mathbf{0.76}$ **m** or **760 mm**

1 atm pressure is sometimes called 760 mm Hg or "760 Torr."

2. The column of water in Figure 10.3 has a height, $h = 0.15$ m. Find the pressure, P.

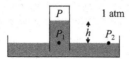

FIGURE 10.3   Column of water of height, h.

$P_1 = P + \rho g h = P + (1000)(9.8)(0.15) = P + 1470$

$P_2 = 1 \text{ atm} = 1.013 \times 10^5 \text{ Pa}$

Set them equal: $P + 1470 = 1.013 \times 10^5$

$$P = 9.98 \times 10^4 \text{ Pa}$$

QUIZ 10.1

1. A submarine is in the water. The water pressure is 7.0 atm. A hatch on the outside of the submarine has an area of 2.3 m². How much force does the water exert on the hatch?
2. In the previous problem, what was the submarine's depth?
3. A U-tube contains water (Figure 10.4). One end is open to the air. The other end contains gas with a pressure 1.020×10^5 Pa. Find the height difference, h.

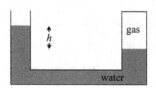

FIGURE 10.4   Water in a U-tube.

4. Suppose the air pressure is $P = 1.005 \times 10^5$ Pa. Find the height, h, of the mercury column to the nearest mm (Figure 10.5).

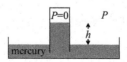

FIGURE 10.5   Column of mercury of height, h.

BUOYANCY

In a liquid, pressure increases with depth. Suppose an object is submerged in the liquid (Figure 10.6). The bottom surface of the object experiences a larger pressure than the top. This pressure difference pushes the object up. The upward force is called *buoyancy*.

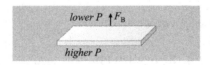

lower P $\uparrow F_B$

higher P

FIGURE 10.6 Object submerged in a liquid.

The magnitude of the buoyant force equals the weight of the displaced fluid,

$$F_B = \rho_f V_f g$$

where ρ_f is the fluid density (kg/m³), and V_f is the volume (m³) of the displaced fluid. The displaced fluid is the fluid that is pushed out of the way because of the object.

1. A rock with a density of 5000 kg/m³ and volume of 0.010 m³ is completely submerged in water. Find the total force on the rock.

   Because the rock is completely underwater, V_f equals the volume of the rock.

   Buoyant force (up): $\rho_f V_f g = (1000)(0.01)(9.8) = 98$ N

   Mass of the rock: $m = (5000 \text{ kg/m}^3)(0.01 \text{ m}^3) = 50$ kg

   Gravity force (down): $mg = (50)(9.8) = 490$ N

   The net force is $98 - 490 = -392$ N (down)

2. A raft with a mass of 40 kg floats on the water (Figure 10.7). It is square with sides 2.0 m and height 0.25 m. Find x, the height that is under water.

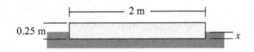

2 m

0.25 m

x

FIGURE 10.7 Raft floating on the water.

Volume of displaced fluid: $V_f = (2)(2)(x) = 4x$

Buoyant force (up): $\rho_f V_f g = (1000)(4x)(9.8) = 39,200x$

Gravity force (down): $mg = (40)(9.8) = 392$

Because the object floats, the buoyant force balances gravity.

$39,200x = 392$

$x = \textbf{0.010 m}$

MOVING FLUIDS

Suppose water is flowing through a pipe. The *volume flow rate* (Q) is the volume of water that flows through the pipe every second. Q has units of m³/s. For a fluid flowing at a speed, v, through a cross-sectional area, A,

$$Q = Av$$

If the pipe narrows or widens, assuming steady flow, Q must remain constant. Each second, the same amount of water flows through the wide section and the narrow section (Figure 10.8). This gives us the *equation of continuity*,

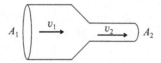

FIGURE 10.8   Fluid flowing through a pipe.

$$A_1 v_1 = A_2 v_2$$

Because 1 m³ is a large volume, sometimes liters (L) are used; $1\,L = 10^{-3}\,m^3$.

1. 7.0 liters of water flow through a pipe of radius 5.0 cm every second. Find the water speed.

   $7\,L = 7 \times 10^{-3}\,m^3$, so $Q = 7 \times 10^{-3}\,m^3/s$

   $A = \pi r^2 = (3.14)(0.05^2) = 7.85 \times 10^{-3}\,m^2$

   $Q = Av$

   $7 \times 10^{-3} = (7.85 \times 10^{-3})v$

   $\textbf{0.89 m/s} = v$

2. Air flows through a 10-cm diameter tube at a speed 4.0 m/s. The tube narrows to a 2.0-cm diameter. What is the air speed in the narrow section?

   $A_1 = \pi(0.05^2) = 0.0025\pi$

   $A_2 = \pi(0.01^2) = 0.0001\pi$

$A_1v_1 = A_2v_2$

$(0.0025\pi)(4) = (0.0001\pi)v_2$

100 m/s $= v_2$

QUIZ 10.2

1. A plastic toy has a density of 950 kg/m³ and a volume of 1.0×10^{-4} m³. It is completely submerged under water. Find the net force on the toy.
2. A 0.60-kg wooden cube has dimensions of 10 cm × 10 cm × 10 cm. It floats on the water. What is the height of the portion of the cube that is under water?
3. Air flows through a 0.25-m diameter duct at 800 L/s. Find the air speed.
4. A wine bottle has a 7.5-cm diameter at the base and 2.5-cm diameter at the opening. Wine pours out the opening at 9.0 cm/s. What is the speed of the wine flow at the base?

BERNOULLI'S EQUATION

Suppose a fluid travels through a pipe (Figure 10.9). We trace the path of the fluid. Bernoulli's equation relates the pressure (P) and speed (v) of the fluid at any two points:

$$P_1 + \tfrac{1}{2}\rho v_1^2 + \rho g y_1 = P_2 + \tfrac{1}{2}\rho v_2^2 + \rho g y_2$$

where y is the height. Note that we ignore viscosity (resistance to flow).

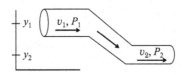

FIGURE 10.9   Fluid traveling through a pipe.

1. Water flows through a wide horizontal tube at a speed $v_1 = 4.0$ m/s (Figure 10.10). The tube narrows, and the water shoots into the air with a speed $v_2 = 16$ m/s. What is the water pressure in the wide section?

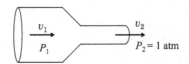

FIGURE 10.10   Water flowing through a horizontal tube.

Because the tube is horizontal, we can set $y_2 = y_1$ and drop from Bernoulli's equation.

$$P_1 + \tfrac{1}{2}\rho v_1^2 = P_2 + \tfrac{1}{2}\rho v_2^2$$

$$P_1 + \tfrac{1}{2}(1000)(4^2) = 1.013 \times 10^5 + \tfrac{1}{2}(1000)(16^2)$$

$$P_1 = \mathbf{2.213 \times 10^5 \ Pa}$$

2. Water is pumped through a pipe up a 10-m high hill. At the bottom of the hill, water flows at 2.0 m/s and a pressure gauge reads 200.5 kPa. At the top, the pressure gauge reads 100.0 kPa. What is the water speed at the top?

   Note that a pressure *gauge* reports the pressure relative to 1 atm, so a reading of 100.0 kPa means the absolute pressure is 100.0 + 101.3 = 201.3 kPa. Because Bernoulli's equation only cares about differences in P, it's okay to use the gauge pressures.

   $$P_1 + \tfrac{1}{2}\rho v_1^2 + \rho g y_1 = P_2 + \tfrac{1}{2}\rho v_2^2 + \rho g y_2$$

   $$2.005 \times 10^5 + \tfrac{1}{2}(1000)(2^2) + 0 = 1.000 \times 10^5 + \tfrac{1}{2}(1000)v_2^2 + (1000)(9.8)(10)$$

   $$1.005 \times 10^5 - 96{,}000 = (500)v_2^2$$

   $$9 = v_2^2, \text{ so } v_2 = \mathbf{3.0 \ m/s}$$

STREAMLINES

Imagine we painted three air molecules black and monitored their paths as they flowed through a tube. They would trace lines called *streamlines*. In Figure 10.11, $v_1 < v_2$. The streamlines are spaced far apart in the wide section, where the velocity is low. From Bernoulli's equation, if velocity is low, then pressure is high. Therefore,

- If the streamlines are far apart, the velocity is low, and the pressure is high.
- If the streamlines are close together, the velocity is high, and the pressure is low.

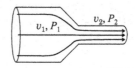

FIGURE 10.11   Streamlines in a horizontal tube.

1. Air (density 1.2 kg/m³) flows through a horizontal tube of cross-sectional area 0.10 m² at a speed 50 m/s and pressure 102.0 kPa. The tube narrows to 0.050 m². What is the pressure in the narrow section?

$$A_1v_1 = A_2v_2$$

$$(0.1)(50) = (0.05)v_2$$

$$100 \text{ m/s} = v_2$$

Because the tube is horizontal, we can set $y_2 = y_1$ and drop from Bernoulli's equation.

$$P_1 + \tfrac{1}{2}\rho v_1^2 = P_2 + \tfrac{1}{2}\rho v_2^2$$

$$1.020 \times 10^5 + \tfrac{1}{2}(1.2)(50^2) = P_2 + \tfrac{1}{2}(1.2)(100^2)$$

$$\mathbf{9.75 \times 10^4 \text{ Pa}} = P_2$$

2. Wind blows over the roof of a house (Figure 10.12). Is the pressure above the roof higher or lower than the pressure inside the house?

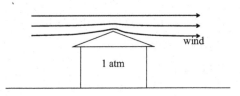

FIGURE 10.12   Wind blowing over a house.

The high air speed of the wind means that the pressure is **lower** than 1 atm.

The pressure difference causes an upward force, which might tear off the roof!

QUIZ 10.3

1. Water flows at 5.0 m/s through a narrow horizontal pipe. A pump maintains a pressure of 102 kPa. The pipe widens. In the wide section, the water speed is 1.0 m/s. What is the pressure in the wide section?
2. Oil (density 900 kg/m³) is poured into a pipe at the top of a 20-m high hill, where the pressure is 1 atm. The oil starts from rest. At the bottom of the hill, the oil flows at 10 m/s. What is the oil pressure at the bottom?
3. Water flows through a horizontal, narrow tube with an area of 0.030 m² at a speed 5.0 m/s. The tube widens to 3.0 m². A pressure gauge in the wide section reads 123.0 kPa. What is the pressure reading in the narrow section?
4. Air flows over a plastic ball. What direction is the force on the ball (up or down)?

CHAPTER SUMMARY

Pressure (N/m², or Pa)	$P = F/A$
Static fluids	$P = P_0 + \rho g d$
	$F_B = \rho_f V_f g$
Moving fluids	$Q = Av$
	$A_1 v_1 = A_2 v_2$
	$P_1 + \frac{1}{2}\rho v_1^2 + \rho g y_1 = P_2 + \frac{1}{2}\rho v_2^2 + \rho g y_2$
Atmospheric pressure (1 atm)	1.013×10^5 Pa $= 101.3$ kPa $= 760$ mm Hg
Density of water	1000 kg/m³

END-OF-CHAPTER QUESTIONS

1. A scuba diver is at a depth of 20 m. His mass is 80 kg.
   a. What is the water pressure? Express your answer to the nearest atm.
   b. Suppose the buoyant force exactly balances the diver's weight. What is the volume of the diver?
2. The weight of a 10.0-kg mass presses down on the surface of oil (density 900 kg/m³). The oil is in a tall, 4.00-cm diameter cylinder.
   a. What is the pressure at the surface of the oil? (Hint: Don't neglect air pressure).
   b. What is the pressure at a depth of 100 cm?
3. A bucket is filled with water (Figure 10.13). There is a small hole 0.20 m below the surface. What is the speed of the water shooting out of the hole? (The water velocity at point 1 is nearly 0.)

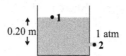

FIGURE 10.13   Water in a bucket.

4. Water flows through garden hose at 10 m/s. A gardener puts a thumb over the end, reducing the cross-sectional area by 50%.
   a. What is the speed of the water that exits the hose?
   b. What is the pressure of the water in the hose?
5. A sprinkler consists of a 1.00-m tall vertical pipe with a head on top. The head has many small holes to let the water out. At the bottom of the pipe, the water pressure is 2.50 atm, and the water flows up at 2.00 m/s. What is the water speed as it exits the head?

ADDITIONAL PROBLEMS

1. A 30-kg ball with a diameter of 0.40 m hangs from a rope. The ball is halfway submerged in water. What is the tension of the rope?

2. "Gauge pressure" is the pressure relative to 1 atm (760 mm Hg). Blood pressure is measured this way. A blood pressure of 120 mm Hg means that the absolute pressure is $120 + 760 = 980$ mm Hg. The blood pressure of a giraffe is 280 mm Hg near its heart. If its head is 2.5 m above the heart, what is its blood pressure there? Assume a blood density of 1000 kg/m^3.

3. Air (density 1.2 kg/m^3) flows through a wide tube with a cross-sectional area of 0.040 m^2 at a speed 10 m/s. The tube narrows to 0.010 m^2. A U-tube, which contains mercury, connects the wide and narrow sections (Figure 10.14). Find the height difference, h.

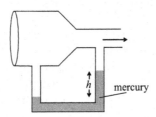

FIGURE 10.14   Air flowing through a horizontal tube, which is connected to a U-tube.

4. Water flows out of a faucet at 1.27 m/s. The opening of the faucet has a diameter of 1.0 cm. The water flows vertically downward.
   a. What is the water speed 10 cm below the faucet opening?
   b. What is the diameter of the water column 10 cm below the faucet opening?

Thermo-dynamics

IDEAL GAS LAW

Most gases obey the ideal gas law,

$$PV = nRT$$

where P is pressure (Pa), V is volume (m³), n is the number of moles (mol), $R = 8.31$ J/(mol·K) is the ideal gas constant, and T is the absolute temperature in Kelvin (K).

The temperature in K is the temperature in °C plus 273. For example, 25°C is $25 + 273 = 298$ K. There are 6.02×10^{23} molecules in 1 mole. One mole has a mass (in grams) equal to the molecular mass. For example, an oxygen atom (O) has an atomic mass of 16. An oxygen molecule (O_2) has a molecular mass $16 + 16 = 32$. One mole of oxygen gas is therefore 32 g.

1. Hydrogen gas is in a 0.50-m³ container. It is held at 27°C and 1 atm. How many moles of gas are there, and what is the mass?

$$PV = nRT$$

$$(1.013 \times 10^5)(0.5) = n\,(8.31)(300)$$

20 mol $= n$

A hydrogen molecule (H_2) has a molecular mass $1 + 1 = 2$, so there are 2 g/mol.

$$\left(\frac{2\text{ g}}{\text{mol}}\right)(20\text{ mol}) = \textbf{40 g}$$

2. In a bottle at 2.0 atm and 0°C is 1.0 mole of oxygen. What is the volume of the bottle in liters?

$$PV = nRT$$

$$(2 \times 1.013 \times 10^5)V = (1)(8.31)(273)$$

$$V = 0.0112 \text{ m}^3$$

$$\left(\frac{1 \text{ L}}{10^{-3} \text{ m}^3}\right)(0.0112 \text{ m}^3) = \textbf{11 L}$$

THERMAL ENERGY

The pressure of a gas results from molecules colliding with the container walls. These collisions exert a force on the walls. When temperature increases, the molecules move faster and have more kinetic energy. The *thermal energy* of an ideal gas is

$$E_{th} = \frac{3}{2} nRT$$

This is the total kinetic energy due to the motion of all the molecules in the gas. (We neglect rotational and vibrational kinetic energy.)

Isothermal means the temperature is constant. *Isobaric* means the pressure is constant.

1. After heating 4.0 moles of gas from 300 to 400 K, what is the change in thermal energy?

$E_{th}(\text{before}) = 3/2 \ (4)(8.31)(300) = 14{,}958 \text{ J}$

$E_{th}(\text{after}) = 3/2 \ (4)(8.31)(400) = 19{,}944 \text{ J}$

$\Delta E_{th} = 19{,}944 - 14{,}958 = \textbf{5000 J}$

2. A gas has a pressure 1.0×10^5 Pa and a volume of 10 m^3. It expands isothermally to 20 m^3. Find the pressure and thermal energy after expansion.

Before: $(10^5)(10) = nRT$

$nRT = 10^6$

Isothermal means T does not change, so nRT is constant.

After: $P(20) = 10^6$

$P = 5 \times 10^4$ **Pa**

$E_{th} = 3/2\ nRT = 1.5 \times 10^6$ **J**

QUIZ 11.1

1. Oxygen gas is in a 2.0-L bottle at 20°C and 1 atm. What is the mass of the oxygen gas?
2. In a container, 1.0 mole of hydrogen gas is at 1 atm and 0°C. What is the volume of the container in liters?
3. After heating 64 grams of oxygen gas from 0°C to 25°C, what is the change in thermal energy?
4. A gas in a 10-L container has a pressure of 0.30 atm. It undergoes isothermal compression to 1.0 L. Find the pressure and thermal energy after compression.

WORK AND HEAT

There are two ways to change the thermal energy of a gas: work or heat. Recall from Chapter 6 that work is force times distance. Suppose you exert a force on a piston and compress a gas. You are doing positive work on the gas.

Assuming constant pressure, the work done on a gas is

$$W = -P\Delta V$$

where ΔV is the volume change. If the gas is compressed, $\Delta V < 0$ and $W > 0$. Positive work is being done *on* the gas. If the gas expands, $\Delta V > 0$ and $W < 0$, this means work is being done *by* the gas (e.g., an expanding piston that drives a crankshaft).

Heat (Q) is energy transferred from a high-temperature substance to a low-temperature one. For example, a flame transfers heat to a pan of water.

1. A gas at 2.0 atm undergoes isobaric expansion. It expands from 3.0 L to 10 L. How much work was done on the gas?

$W = -P\Delta V$

$\quad = - (2 \times 1.013 \times 10^5\ \text{Pa})(7 \times 10^{-3}\ \text{m}^3)$

$\quad = -1400$ **J**

(The work done *by* the gas is $P\Delta V = 1400$ J.)

2. A metal container holds 3.0 moles of gas. The volume of the container is fixed. A professor holds a torch to the container and transfers 300 J of heat to the gas. What is the temperature change of the gas?

Because $\Delta V = 0$, no work is done on the gas. The change in energy is due to heat.

$$\Delta E_{th} = Q = 300 \text{ J}$$

$$\Delta E_{th} = \frac{3}{2} nR\Delta T$$

$$300 = (3/2)(3)(8.31)\,\Delta T$$

8.0 K $= \Delta T$

ENERGY CONSERVATION

The thermal energy of a substance increases when work is done on it (positive W) or when heat is transferred into it (positive Q). The thermal energy decreases when it does work on the outside world (negative W) or when heat flows out (negative Q).

In general, the change in thermal energy is given by the *First Law of Thermodynamics*,

$$\Delta E_{th} = W + Q$$

For an ideal gas,

$$\Delta E_{th} = \frac{3}{2} nR\Delta T$$

1. A machine did 10 J of work compressing a gas. The compression was isothermal. What was the heat flow?

Isothermal means $\Delta T = 0$, so $\Delta E_{th} = 0$

$$\Delta E_{th} = W + Q$$

$$0 = 10 + Q$$

−10 J $= Q$ (heat flowed out of the gas)

2. Helium gas is in a 10-L container. A wall of the container is free to slide. Ice cubes are placed on the container, and the volume decreases to 8.0 L. The pressure, 1 atm, is constant. What is the heat flow?

$$W = -P\Delta V = -(1.013 \times 10^5)(-2 \times 10^{-3}) = 203 \text{ J}$$

$$PV = nRT$$

Before, $nRT = (1.013 \times 10^5)(10 \times 10^{-3}) = 1013$

After, $nRT = (1.013 \times 10^5)(8 \times 10^{-3}) = 810$

$$\Delta E_{th} = \frac{3}{2}(810) - \frac{3}{2}(1013) = -305 \text{ J}$$

$$\Delta E_{th} = W + Q$$

$$-305 = 203 + Q$$

$$\mathbf{-508 \text{ J}} = Q \text{ (heat flows out of the gas)}$$

QUIZ 11.2

1. A constant pressure of 1.3 atm is applied to a gas while it is cooled, causing it to compress from 2.0 L to 1.0 L. How much work was done on the gas?
2. In a container with a fixed volume are 2.0 moles of gas. Ice cubes are placed on the box, causing the temperature to decrease by 5.0°C. What was the heat flow?
3. A gas expands isothermally. The gas does 8.0 J of work. What is the heat flow?
4. An ideal gas expands from 3.0 m³ to 4.0 m³ at a constant pressure of 0.80 atm. What is the heat flow?

HEAT ENGINES

An engine is quite complicated, but here we just look at the basics. Gas is in a cylinder. A heat source such as a flame causes the gas to expand. When it expands, it pushes on a piston, which might turn a crankshaft. Then, the gas is placed in contact with a cold reservoir (e.g., air or water) and it contracts. This expansion-contraction cycle repeats over and over.

During expansion, heat, Q_H, flows from the hot reservoir into the gas (Figure 11.1). During contraction, Q_C flows from the gas to the cold reservoir. Over the entire cycle, the gas does work, W_{out}. For an ideal engine, the work equals the heat that went in minus the heat that went out:

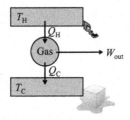

FIGURE 11.1   Diagram of a heat engine.

$$W_{out} = Q_H - Q_C$$

The *efficiency* (*e*) of an engine equals the work divided by how much heat we had to put in:

$$e = \frac{W_{out}}{Q_H}$$

The theoretical maximum efficiency is

$$e_{max} = 1 - T_C/T_H$$

where T_H and T_C are the temperatures of the hot and cold reservoirs measured in K.

1. In a particular heat-engine cycle, 5.0 J of heat go into the gas, and 3.0 J are exhausted. How much work is done in a cycle, and what is the efficiency?

   $W_{out} = Q_H - Q_C = 5 - 3 = \mathbf{2.0\ J}$

   $e = 2/5 = \mathbf{0.40}$ (or 40%)

2. What is the maximum efficiency of a heat engine that uses a flame at 900°C as the hot reservoir and water at 25°C as the cold reservoir?

   $T_H = 900 + 273 = 1173\ K$

   $T_C = 25 + 273 = 298\ K$

   $e_{max} = 1 - T_C/T_H = 1 - 298/1173 = \mathbf{0.75}$ (or 75%)

ENTROPY

If enough heat flows into an ice cube, it will melt, transforming from a solid to a liquid. In a solid, the molecules are fixed in specific locations. In a liquid, the molecules are free to move around. The liquid is more *disordered* than the solid. *Entropy* (S) is a measure of disorder. In this example, $S_{liquid} > S_{solid}$.

In general, if heat Q is put into a substance of temperature T (Kelvin), its entropy will increase by

$$\Delta S = Q/T$$

In this equation, we assume that T is constant while Q is being transferred.

The *Second Law of Thermodynamics* says that, for an isolated system, the total entropy will increase or stay the same: $\Delta S \geq 0$. It turns out that this is what limits the efficiency of heat engines, as defined in the previous section. The law also requires heat to spontaneously flow from a hot object to a cold object and not the other way around.

1. A hot object (380 K) is in contact with a cold object (300 K). From hot to cold, 5.0 J of heat flows. What is the change in entropy of each object? What is the change in entropy of the system?

   ΔS(hot object) = $-5/380 = $ **-0.0132 J/K**

   ΔS(cold object) = $5/300 = $ **0.0167 J/K**

   ΔS(system) = $-0.0132 + 0.0167 = $ **0.0035 J/K**

2. A gas at 300 K expands isothermally and does 50 J of work. What is the change in the entropy of the gas?

   $W = -50$ J

   Isothermal: $\Delta E_{th} = 0$

   $\Delta E_{th} = W + Q$

   $0 = -50 + Q$, or $Q = 50$ J

   $\Delta S = 50/300 = $ **0.17 J/K**

QUIZ 11.3

1. An engine does 100 J of work every second (100 W). Heat goes out the exhaust tailpipe at 80 J per second. What is the efficiency?
2. A hobbyist makes a heat engine, using boiling water (100°C) as the hot reservoir and ice (0°C) as the cold reservoir. What is the maximum possible efficiency?
3. A person with a body temperature of 37°C touches a piece of metal at 23°C. If 0.10 J of heat flows from the person to the metal, find the entropy change of the system (person + metal).
4. A piston does 8.0 J of work on a gas, causing it to compress. The temperature is held fixed at 400 K. What is the change in the entropy of the gas?

CHAPTER SUMMARY

Ideal gas law	$PV = nRT$
	$R = 8.31$ J/(mol·K)
	T (K) $= T$ (°C) $+ 273$
Thermal energy for an ideal gas	$E_{th} = \frac{3}{2}nRT$
Work done on a gas	$W = -P\,\Delta V$
First law of thermodynamics	$\Delta E_{th} = W + Q$
Heat engines	$W_{out} = Q_H - Q_C$
	$e = W_{out}/Q_H$
	$e_{max} = 1 - T_C/T_H$
Change in entropy	$\Delta S = Q/T$

END-OF-CHAPTER QUESTIONS

1. A gas container is filled with 10 moles of helium atoms. The pressure is 1 atm and the temperature is 300 K.
   a. What is the volume of the container?
   b. Find the thermal energy of the gas.
2. In the previous problem, someone puts a flame to the container, raising the gas temperature to 400 K. The pressure remains constant.
   a. What is the change in volume?
   b. What is the change in thermal energy?
   c. How much heat went into the gas?
3. In a metal box with a fixed volume are 2.0 moles of gas. Cold water is spilled on the box, causing the gas temperature to fall from 298 K to 288 K. Find the heat flow.

4. A heat engine has a 200°C hot reservoir and a 40°C cold reservoir.
   a. What is the maximum possible efficiency?
   b. Assuming this efficiency, how much work can the engine do, if 50 J of heat is supplied by the hot reservoir?
5. Water in a puddle has a temperature of 0°C. It slowly freezes, releasing 10 J of heat to the air. The air temperature is −20°C. What is the total change in entropy?

ADDITIONAL PROBLEMS

1. A gas has an initial volume of 12 L, temperature of 300 K, and pressure of 1 atm. The container's top surface is a piston (area 0.010 m²) that can move up or down. Someone places a 20-kg mass on the piston, which compresses the gas. The temperature stays fixed.
   a. What is the volume of the compressed gas?
   b. Did the entropy of the gas increase or decrease?
2. In an insulating container, which does not allow heat to flow in or out, are 4.0 moles of gas. The initial temperature is 300 K. The container expands, doing 1000 J of work on the outside world. What is the temperature of the expanded gas? (This is called *adiabatic* expansion.)
3. At 1 atm and 25°C, a diver inhales 0.25 moles of air with each breath. The diver goes into the water to a depth of 30 m. The temperature of the scuba air is 20°C. How many moles are inhaled with each breath? Assume the volume inhaled is constant.
4. A refrigerator (Figure 11.2) is an example of a *heat pump*. A heat pump is like a heat engine, except that work is done *on* the gas (W_{in}). Q_C is taken from the cold reservoir (e.g., the inside of the refrigerator) and Q_H is put into the hot reservoir (e.g., the kitchen). Suppose a heat pump does 6.0 J of work and expels 14 J of heat into the hot reservoir. How much heat was taken from the cold reservoir?

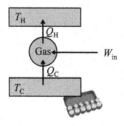

FIGURE 11.2   Diagram of a heat pump.

Electric Fields and Forces

12

THE COULOMB FORCE

An atom consists of a nucleus and electrons. The nucleus has *positive* charge. An electron has *negative* charge. Opposite charges (i.e., positive and negative) attract. Like charges repel.

The unit of charge is the Coulomb (C). An electron has a charge $q = -1.6 \times 10^{-19}$ C. A proton has $q = 1.6 \times 10^{-19}$ C.

The magnitude of the *Coulomb force* of attraction or repulsion between two particles is

$$F = K \frac{|q_1 q_2|}{r^2}$$

where everything is in SI units and $K = 9.0 \times 10^9$.

1. A 4.0×10^{-4} C charge is 2.0 m from a 1.0×10^{-4} C charge. What is the force on the 4.0×10^{-4} C charge?

   $$F = (9 \times 10^9)(4 \times 10^{-4})(1 \times 10^{-4})/2^2 = \mathbf{90 \ N}$$

   Both charges are positive, so they repel each other.
   The direction of the force on the 4.0×10^{-4} C charge is **away from the 1.0×10^{-4} C charge**

2. Find the force on the 1.0 mC charge in Figure 12.1 (mC $= 10^{-3}$ C).

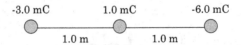

-3.0 mC   1.0 mC   -6.0 mC

          1.0 m   1.0 m

FIGURE 12.1   Three charges. Find the force on the middle charge.

The −3.0 mC charge pulls the 1.0 mC charge toward the left.
The −6.0 mC charge pulls the 1.0 mC charge toward the right.

$$F_x = -(9 \times 10^9)(3 \times 10^{-3})(1 \times 10^{-3})/1^2 + (9 \times 10^9)(6 \times 10^{-3})(1 \times 10^{-3})/1^2$$

$$= 2.7 \times 10^4 \text{ N, toward the right}$$

FORCES AND ANGLES

Suppose we have two positive charges. The force on a charge will be away from the other charge. The same is true for two negative charges.

Now, suppose we have a positive and a negative charge. The force on one charge will be directly toward the other charge.

1. Find the magnitude of the force on the 1.0 μC charge in Figure 12.2 ($\mu C = 10^{-6}$ C).

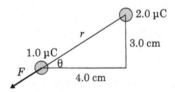

2.0 μC

r

3.0 cm

1.0 μC

F θ

          4.0 cm

FIGURE 12.2   Two charges. Find the force, F.

The distance between the charges is $r = \sqrt{4^2 + 3^2} = 5$ cm $= 0.05$ m

$$F = (9 \times 10^9)(1 \times 10^{-6})(2 \times 10^{-6})/0.05^2 = 7.2 \text{ N}$$

2. In the previous problem, find the components of the force (Figure 12.3).

F

θ F_y

F_x

FIGURE 12.3   Components of the force vector.

$F_x = -(7.2) \cos \theta$ (–because it is being pushed to the left)

$\quad = -(7.2)(4/5) = $ **–5.8 N**

$F_y = -(7.2) \sin \theta$ (–because it is being pushed down)

$\quad = -(7.2)(3/5) = $ **–4.3 N**

QUIZ 12.1

1. A hydrogen atom is an electron and proton separated by 5.3×10^{-11} m. Find the magnitude of the electric force on the electron.
2. Find the force on the charge q, where $q = 1.0\,\mu C$ (Figure 12.4).

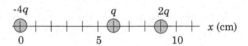

FIGURE 12.4   Three charges. Find the force on the middle charge.

3. Find the magnitude of the force on the $1.0\,\mu C$ charge (Figure 12.5).

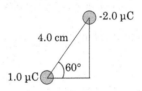

FIGURE 12.5   Two charges. Find the force on the $1.0\,\mu C$ charge.

4. In the previous problem, find the components of the force.

ELECTRIC FIELD

Recall that to calculate the force of gravity on an object, we multiply the mass times **g**. Gravity is an example of a "field." The *electric field*, **E**, is a vector with a magnitude, E. The electric force on a charge, q, is

$$\mathbf{F} = q\mathbf{E}$$

E has units N/C and is called the *electric field strength*. The electric field strength due to a small point charge q is

$$E = K\frac{|q|}{r^2}$$

If q is positive, **E** points away from the charge (Figure 12.6). If q is negative, **E** points toward the charge.

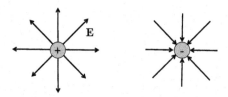

FIGURE 12.6   Electric field from a positive and negative charge.

1. An electric field points straight down. The electric field strength is 10 N/C. Find the force on a particle if its charge is 0.5 C or –0.1 C.

$$E_y = -10 \text{ N/C}$$

For a 0.5 C charge, $F_y = (-10)(0.5) = \mathbf{-5.0 \text{ N/C}}$ (down)

For a –0.1 C charge, $F_y = (-10)(-0.1) = \mathbf{1.0 \text{ N/C}}$ (up)

2. Find the electric field at A (Figure 12.7).

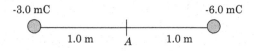

FIGURE 12.7   Two charges. Find the electric field at A.

At A, the electric field due to the –3 mC charge points left, and the electric field due to the –6 mC charge points right.

$E_x = -(9 \times 10^9)(3 \times 10^{-3})/1^2 + (9 \times 10^9)(6 \times 10^{-3})/1^2 = \mathbf{2.7 \times 10^7 \text{ N/C}}$
(toward the right)

ELECTRIC FIELDS AND ANGLES

When adding two electric fields, first find the x and y components of each one. Add the x components to get E_x. Add the y components to get E_y.

1. Find the electric field at A (Figure 12.8).

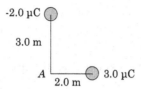

FIGURE 12.8   Two charges. Find the electric field at A.

At A, the electric field from the $-2\ \mu C$ charge points up.

$E_x = 0$

$E_y = (9 \times 10^9)(2 \times 10^{-6})/3^2 = 2{,}000\ \text{N/C}$

The electric field from the $3\ \mu C$ charge points left.

$E_x = -(9 \times 10^9)(3 \times 10^{-6})/2^2 = -6{,}750\ \text{N/C}$

$E_y = 0$

Add the components:

$\mathbf{E_x = 0 - 6{,}750 = -6{,}750\ N/C}$

$\mathbf{E_y = 2{,}000 + 0 = 2{,}000\ N/C}$

2. In the previous problem, what is the electric field strength and direction (Figure 12.9)?

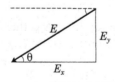

FIGURE 12.9   Components of the electric field.

$E = \sqrt{E_x^2 + E_y^2} = 7{,}040 \text{ N/C}$

$\tan \theta = 2000/6750 = 0.30$

$\theta = \tan^{-1} (0.30) = 17°$

The direction could be expressed as "163° clockwise from the x axis."

QUIZ 12.2

1. A 3.3-ng oil drop has a net charge of 1 electron. What electric field strength will balance its weight?
2. Find the electric field at A (Figure 12.10).

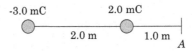

FIGURE 12.10   Two charges. Find the electric field at A.

3. Find the electric field (E_x, E_y) at A (Figure 12.11).

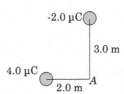

FIGURE 12.11   Two charges. Find the electric field at A.

4. In the previous problem, what is the electric field strength and direction?

ADDING ELECTRIC FIELD VECTORS

Here is an example that puts everything together: adding electric field vectors and then finding the Coulomb force on a test charge.

1. Find the electric field at A (Figure 12.12).

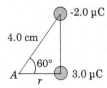

FIGURE 12.12   Two charges. Find the electric field at A.

At *A*, the electric field from the –2 µC charge points up and toward the right.

$$E = (9 \times 10^9)(2 \times 10^{-6})/0.04^2 = 1.125 \times 10^7 \text{ N/C}$$

$$E_x = E \cos (60°) = 5.6 \times 10^6 \text{ N/C}$$

$$E_y = E \sin (60°) = 9.7 \times 10^6 \text{ N/C}$$

The electric field from the 3 µC charge points left.

$$r = 4 \cos (60°) = 2.0 \text{ cm}$$

$$E_x = -(9 \times 10^9)(3 \times 10^{-6})/0.02^2 = -6.75 \times 10^7 \text{ N/C}$$

$$E_y = 0$$

Add the components:

$$E_x = 5.6 \times 10^6 - 6.75 \times 10^7 = \mathbf{-6.2 \times 10^7 \text{ N/C}}$$

$$E_y = 9.7 \times 10^6 + 0 = \mathbf{9.7 \times 10^6 \text{ N/C}}$$

2. In the previous problem, suppose a 10 µC test charge is located at *A*. What is the force on the test charge?

$$\mathbf{F} = q\mathbf{E}$$

$$F_x = (1 \times 10^{-5} \text{ C})(-6.2 \times 10^7 \text{ N/C}) = \mathbf{-620 \text{ N}}$$

$$F_y = (1 \times 10^{-5} \text{ C})(9.7 \times 10^6 \text{ N/C}) = \mathbf{97 \text{ N}}$$

DEFLECTING A CHARGED PARTICLE

An electric field exerts a force on a charged particle. It can be used to alter the trajectory of a moving charged particle.

1. An electron (mass 9.11×10^{-31} kg) travels horizontally at 1.0×10^6 m/s (Figure 12.13). It enters a region between two charged parallel plates where the electric field strength is 900 N/C. Find the electron's acceleration while it is in this region.

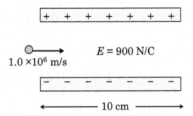

FIGURE 12.13   Electron traveling through parallel plates.

E points away from positive charge and toward negative charge.
Therefore, between the parallel plates, **E** points down.
Unlike point charges, the electric field between parallel plates is constant.

$F_y = qE_y = (-1.6 \times 10^{-19})(-900) = 1.44 \times 10^{-16}$ N

$F_y = ma_y$

$1.44 \times 10^{-16} = (9.11 \times 10^{-31})a_y$

$\mathbf{1.6 \times 10^{14} \ m/s^2} = a_y$

2. In the previous problem, find the electron's final velocity.

$v_x = \mathbf{1.0 \times 10^6 \ m/s}$

Find t, the time the electron is in the parallel-plate region:

$x = v_x t$

$0.10 = (1 \times 10^6)t$

$1 \times 10^{-7} \ s = t$

$v_y = a_y t = (1.6 \times 10^{14})(1 \times 10^{-7}) = \mathbf{1.6 \times 10^7 \ m/s}$

QUIZ 12.3

1. Find the electric field at A (nC = 10^{-9} C) in Figure 12.14.

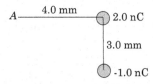

FIGURE 12.14   Two charges. Find the electric field at A.

2. In the previous problem, suppose a 5.0-nC test charge is placed at A. What is the force on the test charge?
3. A proton (mass 1.67 × 10^{-27} kg) travels horizontally at 3.0 × 10^5 m/s. It enters a region between parallel plates where the electric field strength is 10,000 N/C. The positive plate is on top, and the negative plate is on the bottom. Find the proton's acceleration when it is between the plates.
4. In the previous problem, find the proton's velocity after traveling a horizontal distance of 30 cm.

CHAPTER SUMMARY

Coulomb force	$$F = K\frac{|q_1 q_2|}{r^2}$$
	$K = 9.0 \times 10^9$
	Opposite charges attract; like charges repel
Charge of electron	-1.60×10^{-19} C
Charge of proton	1.60×10^{-19} C
Mass of electron	9.11×10^{-31} kg
Mass of proton	1.67×10^{-27} kg
Electric field	$\mathbf{F} = q\mathbf{E}$
Electric field due to point charge	$$E = K\frac{|q|}{r^2}$$
	Positive q: $\mathbf{E}$ points away from charge
	Negative q: $\mathbf{E}$ points toward charge

END-OF-CHAPTER QUESTIONS

1. An amoeba has 1.0 × 10^{16} protons and net charge of 0.30 pC. How many fewer electrons are there than protons? (pC = 10^{-12} C)
2. Calculate the force on charge q, where q = 8.0 nC (Figure 12.15).

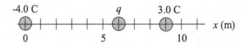

FIGURE 12.15   Three charges. Find the force on the middle charge.

3. Find the position x for which the electric field is 0 (Figure 12.16).

FIGURE 12.16   Two charges. Find where the electric field is 0.

4. Find the electric field strength and direction at A (Figure 12.17). Express the direction as degrees clockwise from the y axis.

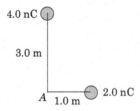

FIGURE 12.17   Two charges. Find the electric field (magnitude and direction) at A.

5. An H_2^+ molecule has two protons and one electron. An electric field points in the $+x$ direction. What is the molecule's acceleration if the electric field strength is 10,000 N/C?

ADDITIONAL PROBLEMS

1. Even if a molecule is neutral, it may be attracted by a charged particle. Consider a *dipole* that consists of two opposite charges ($q = 1.0 \times 10^{-19}$ C) separated by $d = 1.0 \times 10^{-10}$ m (Figure 12.18). Find the force exerted on the dipole.

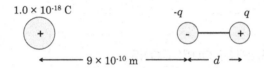

FIGURE 12.18   A positive charge and a dipole.

2. Find the electric field at A, where $Q = 5.0$ nC (Figure 12.19).

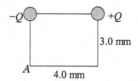

FIGURE 12.19   Two charges. Find the electric field at A.

3. In Figure 12.20, assume the triangle is equilateral. Find the electric field at A.

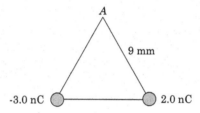

FIGURE 12.20   Two charges. Find the electric field at A.

4. An electron travels at 1.0×10^5 m/s in the $+y$ direction. An electric field (800 N/C) is applied in the $-x$ direction for 3.0×10^{-9} s. Now what is the direction of the velocity? Express as degrees clockwise from the y axis.

Electric Potential

RELATION TO POTENTIAL ENERGY

Recall that an object on a hill has potential energy (PE) due to gravity. When the object slides down the hill, its PE decreases, and its kinetic energy increases.

A particle with charge, q, has PE due to electric forces. The electric *potential* is defined

$$V = PE/q$$

and has units of J/C, called Volts (V).

1. An electron travels from a metal plate at 0 V to a plate at 2.85 V (in vacuum). What is the potential energy change?

   $PE = qV$

   $PE(\text{before}) = (-1.6 \times 10^{-19})(0) = 0$

   $PE(\text{after}) = (-1.6 \times 10^{-19})(2.85) = -4.56 \times 10^{-19}\,J$

   $\Delta PE = \mathbf{-4.56 \times 10^{-19}\,J}$

2. In the previous problem, what was the electron's speed just before it hit the 2.85-V plate? Assume the electron starts from rest.

   From energy conservation, its kinetic energy (KE) equals $-\Delta PE$.

$KE = \frac{1}{2}\,mv^2$

$4.56 \times 10^{-19} = (0.5)(9.11 \times 10^{-31})v^2$

$1.0 \times 10^{12} = v^2$

$\mathbf{1.0 \times 10^6\,m/s} = v$

UNIFORM ELECTRIC FIELD

If the electric field is constant (uniform), the magnitude of the potential change is

$$\Delta V = E{\cdot}d$$

where d is the distance along the **E** field direction. **E** points from high V to low V.
    A particle that travels in the same direction as the **E** vector will experience a drop in its electric potential. A particle that goes opposite to **E** will experience a potential increase.

1. Two charged parallel plates are separated by $d = 10$ cm (Figure 13.1). The plates have a potential difference of 9.0 V. Find the electric field between the plates.

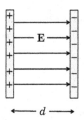

FIGURE 13.1   Electric field between charged parallel plates.

Electric field strength: $E = \Delta V/d = 9.0/0.10 = \mathbf{90\ V/m}$

Direction: **E points from the positive plate to the negative plate**

2. A negatively charged cloud is 700 m above the ground. The ground has positive charge. The electric field strength in the region between the cloud and ground is 1.0×10^6 V/m. What is the electric potential difference between the cloud and ground?

The electric field points from the ground to the cloud.

Therefore, the **ground has higher potential than the cloud**.

$\Delta V = E{\cdot}d = (1 \times 10^6)(700) = \mathbf{7.0 \times 10^8\,V}$

QUIZ 13.1

1. A proton travels from a plate at 48 V to a plate at 0 V (in vacuum). What is the potential energy change?
2. In the previous problem, what was the proton's speed just before it hit the 0-V plate?
3. A 1.5-V battery is attached to two parallel plates (in vacuum) separated by $d = 1.0$ mm (Figure 13.2). The plate attached to the positive (+) terminal of the battery has a potential 1.5 V higher than the other plate. Find the electric field strength between the plates.

FIGURE 13.2   Battery attached to parallel plates.

4. Two charged plates are separated by 1.0 cm. Between them, the electric field strength is 400 V/m. The positively charged plate has a potential 8.5 V. What is the potential of the negative plate?

POTENTIAL DUE TO A POINT CHARGE

The electric potential due to a small particle of charge, q, (point charge) is

$$V = K\frac{q}{r}$$

where r is the distance from the particle. This equation is also valid for a spherical object, where r is the distance from the center of the sphere to a point outside the sphere.

1. A ball of radius 90 cm has a charge of -0.50 µC. What is the potential energy of an electron on its surface?

$V = (9 \times 10^9)(-5 \times 10^{-7})/0.9 = -5 \times 10^3$ V

$PE = qV = (-1.6 \times 10^{-19})(-5 \times 10^3)$

$= 8.0 \times 10^{-16}$ J

2. In the previous problem, the electron accelerates away from the surface. What is its speed when it is far away?

When r is very large, V is practically 0 because r is in the denominator.

Therefore, when the electron is far away, its PE = 0.

PE(before) = 8.0×10^{-16} J

PE(after) = 0

From energy conservation, KE = $-\Delta$PE = 8.0×10^{-16} J

$\frac{1}{2}(9.11 \times 10^{-31})v^2 = 8 \times 10^{-16}$

$v^2 = 1.76 \times 10^{15}$

$v = \mathbf{4.2 \times 10^7\, m/s}$

POTENTIAL DUE TO SEVERAL POINT CHARGES

To calculate the electric potential due to several point charges, we just add the potentials due to each charge.

1. Find the value of x where $V = 0$ (Figure 13.3). Let x be between the charges.

-3.0 mC   6.0 mC x (m)
0   1   2

FIGURE 13.3   Two charges. Find the point between the charges where $V = 0$.

The distance from the −3.0 mC to x is just x.

The distance from the 6.0 mC charge to x is $2 - x$.

$V = K(-3 \times 10^{-3})/x + K\,(6 \times 10^{-3})/(2 - x) = 0$

$-3/x + 6/(2 - x) = 0$

$3/x = 6/(2 - x)$

$3(2 - x) = 6x$

$6 = 9x \rightarrow x = \mathbf{0.67\ m}$

2. Find the potential at A (Figure 13.4).

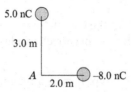

FIGURE 13.4 Two charges. Find the potential at A.

$V = (9 \times 10^9)(5 \times 10^{-9})/3 + (9 \times 10^9)(-8 \times 10^{-9})/2$

$= 45/3 - 72/2 = \mathbf{-21\ V}$

QUIZ 13.2

1. A sphere of radius 1.0 m has a charge of 10 µC. A proton is on the surface, initially at rest. What is the proton's potential energy?
2. In the previous problem, the proton leaves the surface of the sphere. What is the speed of the proton when it is far away?
3. Find the value of d for which the electric potential is 0 (Figure 13.5).

FIGURE 13.5 Two charges. Find d such that $V = 0$.

4. Find the potential at A (Figure 13.6).

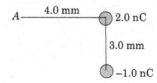

FIGURE 13.6 Two charges. Find the potential at A.

CONDUCTORS

Inside a conductor (e.g., metal), electrons are free to move. A conductor has the following properties:

- **E** = 0 inside the conductor.
- If there is an **E** field outside the conductor, **E** is perpendicular to the surface.
- V = constant throughout the conductor.
- Excess charge (positive or negative) resides on the surface of the conductor.

1. A sharp metal tip points up. It is in a region where the electric field points down. Draw the electric field lines.

   The electric field lines close to the tip bend toward the surface (Figure 13.7). Notice that, near the tip, the lines are closely spaced. That means the electric field strength is high in that region.

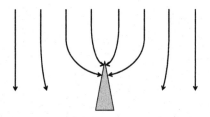

FIGURE 13.7   Electric field near a sharp metal tip.

2. In a region of space, the **E** field points toward the right. A metal sphere is placed in this region. Draw the electric field lines and charge on the metal.

   Far from the metal, the electric field lines are horizontal (Figure 13.8). Near the metal surface, they bend to be perpendicular to the surface.

   Inside the conductor, **E** pushed electrons to the left, leaving behind positive charge. The negative and positive charges produced an electric field, which exactly canceled **E** inside the conductor.

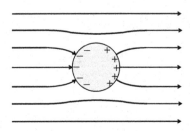

FIGURE 13.8   Electric field near a metal sphere.

INSULATORS AND CAPACITORS

Examples of insulators are glass or plastic. Charges can accumulate on the surface of an insulator. Unlike a conductor, electrons cannot travel through an insulator.

A parallel-plate *capacitor* consists of an insulator sandwiched between two metal plates. Suppose a battery is attached to the plates. The positive terminal will provide a charge $+Q$ to a plate, and the negative terminal will provide $-Q$ to the other plate. The potential difference between the plates is ΔV. For a capacitor,

$$Q = C\Delta V$$

where C is the *capacitance*, measured in units of Farads (F). For a parallel-plate capacitor,

$$C = \kappa\varepsilon_0 A/d$$

where κ is the dielectric constant (a property of the insulator), $\varepsilon_0 = 8.85 \times 10^{-12}$, A is the area of the plates (m²), and d is the distance between the plates (m).

1. A 1.5-V battery is attached to a 5 pF parallel-plate capacitor ($pF = 10^{-12}$ F). One plate is at 0 V, or "ground." Find the charge on each plate, and sketch the equipotential (constant voltage) lines.

   The sketch is shown in Figure 13.9.

   $Q = (5 \times 10^{-12})(1.5) = \mathbf{7.5 \times 10^{-12}\,C}$

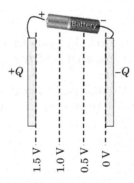

FIGURE 13.9   Battery attached to parallel plates. Equipotential lines are shown.

2. A student makes a capacitor from a 1.0-cm × 1.0-cm × 0.5-mm piece of glass ($\kappa = 4.8$). Silver is painted on each 1-cm² side and a 9.0-V battery is attached. How much charge is on each side?

   $C = (4.8)(8.85 \times 10^{-12})(0.01^2)/(5 \times 10^{-4}) = 8.5 \times 10^{-12}\,F$

   $Q = C\Delta V = (8.5 \times 10^{-12})(9) = \mathbf{7.7 \times 10^{-11}\,C}$ (one plate has Q, the other plate has $-Q$).

QUIZ 13.3

1. A square piece of aluminum (Figure 13.10) is placed in a region where the electric field points up. Sketch the electric field lines.

FIGURE 13.10   Piece of aluminum.

2. A gold rugby ball (Figure 13.11) is placed in a region where the electric field points down. Sketch the electric field lines and the charges on the ball.

FIGURE 13.11   Solid gold rugby ball.

3. A 12-V battery is attached to a 15 μF parallel-plate capacitor. The negative battery terminal is grounded. Find the charge on each plate and sketch the equipotential lines.
4. A capacitor consists of 0.10-mm thick Teflon ($\kappa = 2.1$) sandwiched by two 1.0-cm² metal plates; 24 V is applied. How much charge is on each plate?

CHAPTER SUMMARY

Electric potential	$V = PE/q$
Potential change in a constant electric field	$\Delta V = E \cdot d$ (magnitude)
	E points from high V to low V
Potential due to point charge	$V = K\dfrac{q}{r}$
Charges on capacitor plates ($\pm Q$)	$Q = C\Delta V$
Capacitance of a parallel-plate capacitor	$C = \kappa e_0 A/d$
	$\varepsilon_0 = 8.85 \times 10^{-12}$

END-OF-CHAPTER QUESTIONS

1. On a stormy day, a cloud is 500 m above the ground. The cloud has a charge of −640 C and the ground has a charge of 640 C. The electric field strength in the region between the cloud and ground is 8.0×10^5 V/m.
   a. How many excess electrons does the cloud have?
   b. At the ground, $V = 0$. What is the potential of the cloud?
2. An electron is initially at rest and is 0.14 m away from a fixed −0.50-nC point charge. The electron accelerates. How fast is the electron moving when it is 0.20 m from the fixed point charge?
3. Find the values of x where the potential is 0 (Figure 13.12).

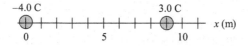

FIGURE 13.12   Two charges. Find where $V = 0$.

4. Find the potential at A (Figure 13.13).

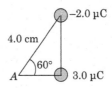

FIGURE 13.13   Two charges. Find the potential at A.

5. A 12-V battery is attached to two metal plates (20-cm × 40-cm) separated by a 1.0-mm thick piece of glass ($\kappa = 11.3$). One metal plate is grounded.
   a. Calculate the charge on each plate.
   b. Sketch the parallel plates with electric field and equipotential lines.

ADDITIONAL PROBLEMS

1. Two protons are initially at rest and a distance of 9.0×10^{-10} m apart. Find their speed when they are far apart.
2. Find the value of x between the two charges where $V = 9.0$ V (Figure 13.14).

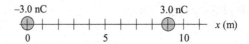

FIGURE 13.14 Two charges. Find the point between the charges where $V = 9.0$ V.

3. Find the points on the y axis where the potential is 0 (Figure 13.15).

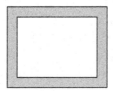

FIGURE 13.15 Two charges. Find the points on the y axis where $V = 0$.

4. A "Faraday cage" can be made from a hollow metal box (Figure 13.16). Suppose the box is placed in a region where the electric field points up. Sketch the electric-field lines and indicate the value of the **E** field inside the box.

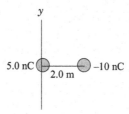

FIGURE 13.16 Hollow metal box.

Electrical Circuits

CURRENT AND OHM'S LAW

Current (I) is the flow of charge, defined as the number of Coulombs flowing past a particular point each second. The units are C/s, commonly called amps (A).

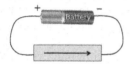

FIGURE 14.1 Battery attached to a material. The arrow shows the current direction.

If we attach a voltage source (e.g., battery) to an insulator, very little current will flow. If we attach it to a conductor, the current will be large. In general, current can be determined from *Ohm's law*,

$$\Delta V = IR$$

where ΔV is the potential difference, and R is the material's resistance in Ohms (Ω). Current flows from high to low potential (Figure 14.1).

Note that, because electrons are negative, they actually move opposite to the current direction. Positive particles (e.g., positive ions) move in the same direction as the current.

1. A current of 3.2 mA flows through a wire. How many electrons per second is this?

    $(3.2 \times 10^{-3}\,\text{C/s})(1\ \text{electron}/1.6 \times 10^{-19}\,\text{C})$
    $= \mathbf{2.0 \times 10^{16}\ electrons/s}$

2. A 100-Ω resistor is attached to a 1.5-V battery. Find the current.

$\Delta V = IR$

$1.5 = I\,(100)$

0.015 A $= I$ (or, $I = 15$ mA)

POWER

Recall that power = force $\times$ velocity and is measured in Watts (W). A battery maintains a voltage (potential difference), ΔV. This potential difference provides the force that pushes charged particles, resulting in a current, I. The power provided by the battery is

$$P = I\,\Delta V$$

A resistor that has current flowing through it has a voltage drop (i.e., the current flows from high voltage to low voltage). If the voltage drop is ΔV, then the power dissipated by the resistor is $P = I\,\Delta V$. The power is dissipated as heat.

1. A 12-V battery provides 3.0 A to a load. How much power does the battery provide?

$P = (3)(12) = $ **36 W**

2. A 9.0-V battery is connected to a 100-Ω resistor. What is the resistor's power dissipation?

First, use Ohm's law to find I.

$\Delta V = IR$

$9 = I\,(100)$

$0.09 = I$

Second, use $P = I\,\Delta V$.

$P = (0.09)(9) = $ **0.81 W**

QUIZ 14.1

1. A lightning bolt delivers a current of 30,000 A for 0.50 ms. How many electrons are delivered?
2. A person puts one hand on the negative terminal and the other hand on the positive terminal of a 12-V battery. How much current flows through the person? The person's resistance is 100 kΩ.

3. A generator provides 100 V at a maximum current of 5.0 A. What is the maximum power that the generator can provide?

4. A small light bulb has a resistance of 0.50 Ω. It is attached to two 1.5-V batteries in series (together, they provide 3.0 V). How much power does the light bulb dissipate?

RESISTORS IN SERIES

If resistors are in series, the same current I flows through each resistor. The voltage drop across each resistor is given by Ohm's law, $\Delta V = IR$. Figure 14.2 shows a circuit with a battery (V) and two resistors (R_1 and R_2) in series. The positive terminal of the battery is indicated. To find the current, we *add* the resistors,

$$R = R_1 + R_2$$

and then use Ohm's law to solve for I.

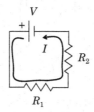

FIGURE 14.2   Two resistors and a voltage source (battery) in series.

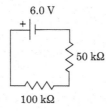

FIGURE 14.3   A series circuit.

1. Find the current (Figure 14.3).

$R = 50 + 100 = 150 \text{ k}\Omega$

$\Delta V = IR$

$6 = I (1.5 \times 10^5)$

$\mathbf{4.0 \times 10^{-5} \text{ A}} = I$

2. In the previous problem, find the voltage drop across each resistor.

100-kΩ resistor: $\Delta V = (4 \times 10^{-5})(1 \times 10^5) = $ **4.0 V**

50-kΩ resistor: $\Delta V = (4 \times 10^{-5})(5 \times 10^4) = $ **2.0 V**

Notice how the voltage drops add up to 6.0 V, which is the voltage provided by the battery.

RESISTORS IN PARALLEL

If resistors are in parallel (Figure 14.4), each resistor has the same voltage drop. The current flowing through a resistor is given by Ohm's law, $\Delta V = IR$.

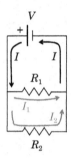

FIGURE 14.4 Two resistors in parallel.

To find the current, I, flowing into two resistors in parallel, first calculate an equivalent resistance using

$$1/R = 1/R_1 + 1/R_2$$

and then use Ohm's law. (For more than two resistors, we have $1/R_1 + 1/R_2 + 1/R_3 + ...$)

1. Find the current provided by the voltage source (Figure 14.5).

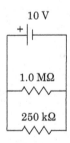

FIGURE 14.5 A parallel circuit.

$$1/R = 1/(1 \times 10^6) + 1/(0.25 \times 10^6) = 5 \times 10^{-6}$$

$$R = 1/(5 \times 10^{-6}) = 2 \times 10^5 \ \Omega$$

$$\Delta V = IR$$

$$10 = I \, (2 \times 10^5)$$

$$\mathbf{5.0 \times 10^{-5} \, A} = I$$

2. In the previous problem, find the current flowing through each resistor.

1.0-MΩ resistor: $10 = I_1(1 \times 10^6)$

$$\mathbf{1.0 \times 10^{-5} \, A} = I_1$$

250-kΩ resistor: $10 = I_2(0.25 \times 10^6)$

$$\mathbf{4.0 \times 10^{-5} \, A} = I_2$$

Notice how the two currents add up to the total current $I = 5.0 \times 10^{-5} \, A$.

QUIZ 14.2

1. A circuit consists of a 9.0-V battery, two 100-Ω resistors, and one 50-Ω resistor, in series. Find the current.
2. In the previous problem, calculate the voltage drop across each resistor.
3. Find the current provided by the battery (Figure 14.6).

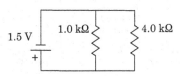

FIGURE 14.6   Another parallel circuit.

4. In the previous problem, calculate the current flowing through each resistor.

KIRCHOFF'S LOOP RULE

The loop rule helps us solve for voltage drops or currents. First, draw a current loop in the circuit. (Don't worry if you get the direction wrong; if you do, I will be negative.) Going around the loop, we add the voltage changes (rises or drops). The sum of the voltage changes must equal 0.

If the loop goes in the same direction as the current, a resistor has a voltage drop. Otherwise, it's a voltage rise. For a voltage source, if the current goes from – to +, the voltage change is positive; otherwise it's negative.

1. Find the current (Figure 14.7).

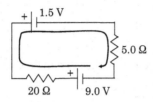

FIGURE 14.7   Circuit with a current loop drawn.

Going around the loop, we have:

$9 - I\,(20) - 1.5 - I\,(5) = 0$

$7.5 - 25I = 0$

$25I = 7.5$

$I = 0.30\ \text{A}$

2. Find the current flowing through the resistor when the switch is closed, and the voltage drop across the switch when it is open (Figure 14.8).

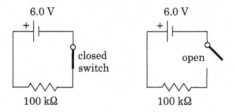

FIGURE 14.8   Circuit with a closed switch and an open switch.

The closed switch has 0 resistance.

$6 - I\,(100{,}000) - I\,(0) = 0$

$6 = I\,(100{,}000)$

$6.0 \times 10^{-5}\,\text{A} = I$

The open switch has infinite resistance, so current is 0.

Let ΔV_{switch} = the voltage drop across the switch.

$6 - (0)(100,000) - \Delta V_{\text{switch}} = 0$

$\mathbf{6.0\ V} = \Delta V_{\text{switch}}$

KIRCHOFF'S CURRENT RULE

Consider two wires that were soldered at a single point or junction (Figure 14.9). The current flowing *into* the junction must equal the current flowing *out of* the junction.

1. Find the current supplied by the voltage source (I) and the current flowing through each resistor (I_1 and I_2) in Figure 14.10.

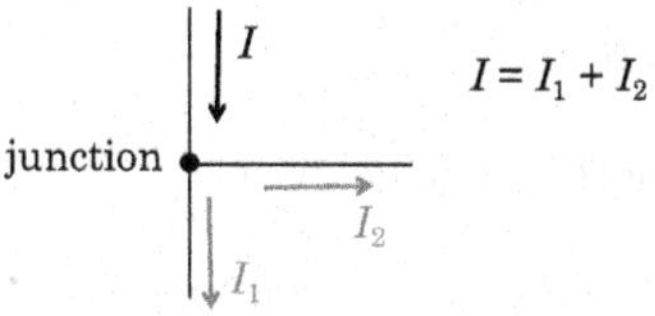

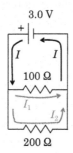

FIGURE 14.9   Current flowing into (I) and out of (I_1, I_2) a junction.

3.0 V

100 Ω

200 Ω

FIGURE 14.10   Currents in a parallel circuit.

Use Ohm's law to find the currents through each resistor:

$3 = I_1 (100)$

$\mathbf{0.030\ A} = I_1$

$3 = I_2 (200)$

$\mathbf{0.015\ A} = I_2$

Kirchoff's junction rule: $I = 0.030 + 0.015 = \mathbf{0.045\ A}$

2. Find the current flowing through each resistor (Figure 14.11).

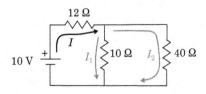

12 Ω

10 V

10 Ω   40 Ω

FIGURE 14.11   Currents in a circuit.

Use the Kirchoff loop rule for the leftmost loop, letting $I = I_1 + I_2$.

$10 - (I_1 + I_2)12 - I_1(10) = 0$

$10 - 22\,I_1 - 12\,I_2 = 0$

Now do the rightmost loop (clockwise):

$-I_2(40) + I_1(10) = 0$ (+ because I_1 goes opposite to the clockwise path)

$I_2 = I_1/4$

Plug this into the underlined equation (1).

$10 - 22\,I_1 - 2\,I_2 = 0$ as (1)

$10 = 25\,I_1$

$I_1 = \mathbf{0.40\ A}$

$I_2 = I_1/4 = \mathbf{0.10\ A}$

$I = I_1 + I_2 = \mathbf{0.50\ A}$

QUIZ 14.3

1. Find the current in Figure 14.12.

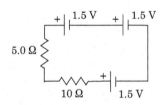

1.5 V   1.5 V

5.0 Ω

10 Ω   1.5 V

FIGURE 14.12   A complicated series circuit.

2. Find the voltage drop across the 90-kΩ resistor in Figures 14.13a and b.

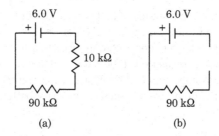

(a)   (b)

FIGURE 14.13   Two circuits.

3. Find the current flowing through each resistor, and the current supplied by the voltage source (Figure 14.14).

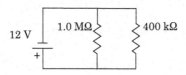

FIGURE 14.14   A parallel circuit.

4. Find the current flowing through each resistor in Figure 14.15.

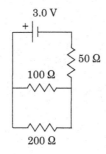

FIGURE 14.15   A more complicated circuit.

CHAPTER SUMMARY

Ohm's law	$\Delta V = IR$
Power	$P = I\,\Delta V$
Resistors in series	$R = R_1 + R_2 + \ldots$
Resistors in parallel	$1/R = 1/R_1 + 1/R_2 + \ldots$
Kirchoff's loop rule	Sum of voltage changes $= 0$
Kirchoff's current rule	Current flowing into junction $=$ current flowing out of junction

END-OF-CHAPTER QUESTIONS

1. Fred's tongue has a resistance of 50 kΩ. He attaches a 9.0-V battery to it. Find the current flow and power dissipation.
2. Two identical resistors are connected in series to a 100-V voltage source. The current flow is measured to be 5.0 mA. How much power will be supplied by the voltage source if the resistors are connected in parallel?
3. Find the current and power supplied by the 9.0-V battery (Figure 14.16).

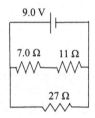

FIGURE 14.16   Circuit powered by a 9.0-V battery.

4. Find the following for the 3.0-Ω resistor (Figure 14.17):

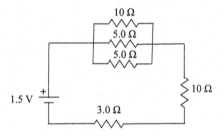

FIGURE 14.17   Solve for current, voltage, and power dissipation of the 3.0-Ω resistor.

   a.   Current flow
   b.   Voltage drop
   c.   Power dissipation
5. Find the current flowing through each of the resistors in Figure 14.18.

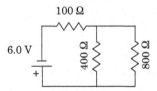

FIGURE 14.18   Find the currents.

ADDITIONAL PROBLEMS

1. The problems in this chapter have been for direct-current (DC) voltage sources, which are constant. A wall plug is an example of an alternating-current (AC) source, which supplies a voltage that is sinusoidal. Let $V = (120 \text{ V}) \sin (380t)$. Find the average power dissipated by the resistor (Figure 14.19). (Hint: a $\sin^2$ or $\cos^2$ function oscillates between 0 and 1, for an average value of ½.)

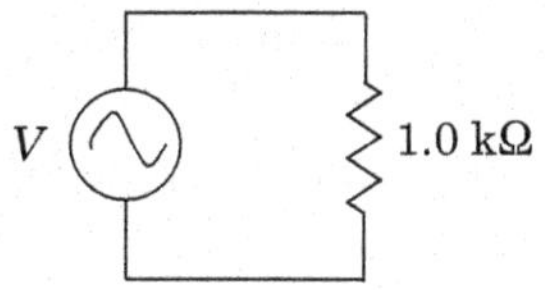

V ⊗   1.0 kΩ

FIGURE 14.19   Alternating-current (AC) circuit. The sine-wave symbol means the voltage supply varies over time.

2. Find the current when the switch is closed and open (Figure 14.20).

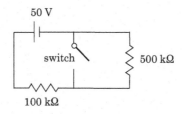

50 V

switch   500 kΩ

100 kΩ

FIGURE 14.20   Circuit with a switch (shown open).

3. Find the current flowing through each resistor (Figure 14.21).

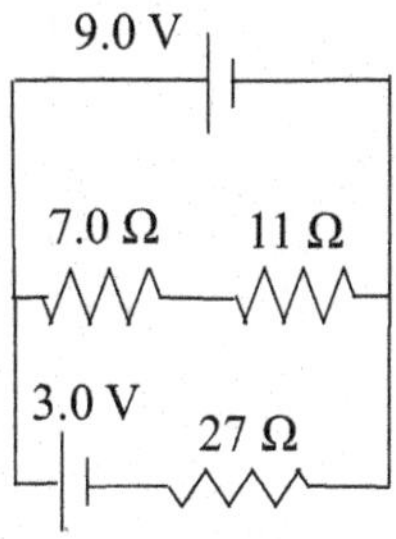

9.0 V

7.0 Ω   11 Ω

3.0 V   27 Ω

FIGURE 14.21   Find the currents.

4. Find the current supplied by the 3.0-V battery (Figure 14.22).

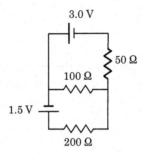

FIGURE 14.22   Find the current supplied by the 3.0-V battery.

Magnetic Fields and Forces

15

MAGNETS

A magnet has a north (N) and south (S) pole. Similar to electric charges, the same poles (N and N, S and S) repel, and opposite poles (N and S) attract.

A magnet produces a *magnetic field*, **B**, with units of Tesla (T). Like an electric field, a magnetic field is a vector. Outside a magnet, the **B** field loops from N to S (Figure 15.1). Inside the magnet, the **B** field goes from S to N, completing the loop.

Suppose a magnet is placed in a magnetic field produced by the Earth or some other external source. The N pole will be pushed in the direction of **B**, and the S pole will be pushed opposite to **B**. (N and S are analogous to + and − charges.) This causes the magnet to align itself parallel to the magnetic field.

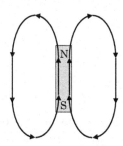

B field due to magnet

Magnet lines up with external **B** field

FIGURE 15.1 A magnet produces a **B** field and aligns parallel to an external **B** field.

1. Which orientation of the bar magnet will be attracted to the horseshoe magnet (Figure 15.2)?

(a)   (b)

FIGURE 15.2 Horseshoe magnet and bar magnet.

The answer is **(b)** because N and S attract each other.

2. Magnet #1 is fixed and Magnet #2 is free to rotate (Figure 15.3). Find θ after Magnet #2 comes to rest.

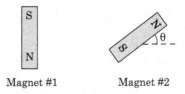

Magnet #1   Magnet #2

FIGURE 15.3 Two magnets. Magnet #2 is free to rotate.

The **B** field to the right of Magnet #1 points up.

Magnet #2 will align with the **B** field, so $\theta = 90°$

FERROMAGNETISM

Iron, nickel, and cobalt are *ferromagnetic*. A ferromagnetic sample (e.g., a steel nail) has many magnetic domains, which act like randomly oriented bar magnets (Figure 15.4). The random orientations cancel each other out, so the sample as a whole is not magnetic. In the presence of an external **B** field, however, the domains line up, and the sample becomes a magnet. If the sample continues to be a magnet after the external **B** field is removed, it is called a *permanent magnet*.

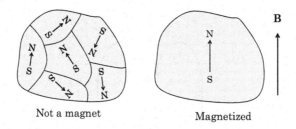

Not a magnet   Magnetized

FIGURE 15.4   A ferromagnetic material can be turned into a magnet by applying an external **B** field.

1. A horseshoe magnet attracts a piece of iron. Draw the **B** field from the horseshoe magnet and the magnetic poles of the iron.
   Outside the horseshoe magnet, the **B** field points from N to S (Figure 15.5). Inside the magnet, the field goes from S to N, completing the loops. The iron piece becomes magnetized. The N and S poles of the two magnets attract each other.

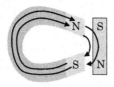

FIGURE 15.5   **B** field lines due to the horseshoe magnet. The **B** field magnetizes the iron.

2. A steel nail is above the N pole of a bar magnet. The nail and magnet are oriented parallel to each other. Draw the **B** field from the bar magnet and the magnetic poles of the nail.

   Above the N pole of the bar magnet, the **B** field points up and magnetizes the nail (Figure 15.6). The S pole of the nail is attracted to the N pole of the magnet.

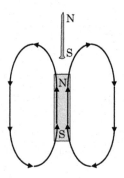

FIGURE 15.6 **B** field lines due to the bar magnet. The **B** field magnetizes the nail.

QUIZ 15.1

1. For which orientation, (a) or (b), will the bar magnets (Figure 15.7) repel each other?

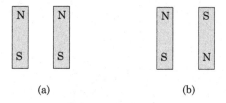

(a)   (b)

FIGURE 15.7 Two orientations of bar magnets.

2. A fixed bar magnet is oriented perpendicular to the page, with S pointing out of the page (Figure 15.8). A second bar magnet is slightly above the fixed magnet and is free to rotate. Draw the **B** field due to the fixed magnet and the orientation of the second magnet.

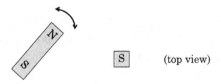

(top view)

FIGURE 15.8 Two bar magnets. One is free to rotate. The second is perpendicular to the page, with S pointing out of the page.

3. A horseshoe magnet attracts a steel sphere. Draw the **B** field from the horseshoe magnet and the magnetic poles of the sphere.

4. A steel nail is magnetized with the north pole at the head. A tiny iron ball is placed on the head of the nail. Draw the **B** field from the nail and the magnetic poles of the ball.

AMPERE'S LAW

Ampere's Law says that electrical current produces a magnetic field. For current in a straight wire, the **B** field follows the *right-hand rule*, where the thumb points in the direction of I (Figure 15.9). The fingers curl in the direction of the magnetic field loops.

For a long, straight wire, the magnitude of the field is

FIGURE 15.9   Right-hand rule for a current I.

$$B = \frac{\mu_0 I}{2\pi r}$$

where $\mu_0 = 4\pi \times 10^{-7}$, r is the distance to the wire, and everything is in SI units. As shown in the figure, **B** is tangential to a circle with the wire in the center.

1. Find the **B** field at point A (Figure 15.10).

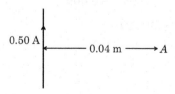

FIGURE 15.10   Current of 0.50 A flowing through a wire.

$B = (4\pi \times 10^{-7})(0.5)/(2\pi \times 0.04) = \mathbf{2.5 \times 10^{-6}\,T}$

Direction: From the right-hand rule, **into the page**

2. In Figure 15.11, two wires have 60 A running through them. Find the **B** field at point A.

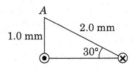

FIGURE 15.11   Two wires with current. The dot means the current runs out of the page, and the cross is into the page.

We will calculate **B** due to each wire, then add them together (Figure 15.12).

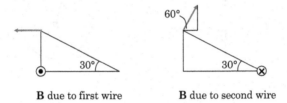

B due to first wire   **B due to second wire**

FIGURE 15.12   The **B** field is perpendicular to the line drawn from the wire to A.

First wire: $B = (4\pi \times 10^{-7})(60)/(2\pi \times 0.001) = 0.012$ T

$B_x = -0.012$ T

$B_y = 0$

Second wire: $B = (4\pi \times 10^{-7})(60)/(2\pi \times 0.002) = 0.006$ T

$B_x = 0.006 \cos(60°) = 0.003$ T

$B_y = 0.006 \sin(60°) = 0.0052$ T

Total: $B_x = -0.012 + 0.003 = \mathbf{-0.009\ T}$

$B_y = 0 + 0.0052 = \mathbf{0.0052\ T}$

CURRENT LOOPS

For current going around a circular loop, the right-hand rule tells us the direction of the **B** field (Figure 15.13). Notice how the field lines curl around, similar to a bar magnet. At the *center* of a single loop of radius, R, the magnitude is given by

$$B = \frac{\mu_0 I}{2R}$$

A *solenoid* is made of many loops. For n loops per meter, the field inside the solenoid is

$$B = \mu_0 n I$$

Note that if the wire is looped around a magnetic metal such as iron, then B can be increased by many times.

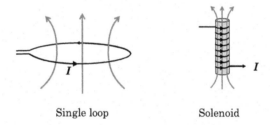

Single loop   Solenoid

FIGURE 15.13   **B** field due to a single loop of wire and a solenoid.

1. Current (2.5 A) flows clockwise through a 3.0-cm diameter loop of wire. Find the **B** field at the center.

$B = (4\pi \times 10^{-7})(2.5)/0.03 = \mathbf{1.0 \times 10^{-4}\,T}$

Direction: **Into the plane** of the loop

2. Suppose the solenoid in Figure 15.13 has a height of 2.0 cm and $I = 2.0$ A. The loops are wound around a nonmagnetic cylinder like plastic. Find the magnitude of the magnetic field inside the solenoid and the orientation of the N and S poles.

$n = 8 \text{ loops}/0.02 \text{ m} = 400 \text{ m}^{-1}$

$B = (4\pi \times 10^{-7})(400)(2) = \mathbf{1.0 \times 10^{-3}\,T}$

From the direction of the magnetic field loops, **N is at the top of the solenoid, S is at the bottom**

QUIZ 15.2

1. An electrician measures the magnetic field 1.0 mm from a wire to be $8.6 \times 10^{-5}\,T$. (This is the amount due to the wire, not the Earth's magnetic field.) How much current is flowing through the wire?
2. Two parallel wires have currents of 15.0 A, flowing in opposite directions (Figure 15.14). Find the magnetic field at A.

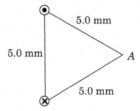

5.0 mm

5.0 mm

A

5.0 mm

FIGURE 15.14   Two wires with current. The dot means the current runs out of the page, and the cross is into the page.

3. A student has a loop of wire with a radius of 4.0 mm and wants to produce a magnetic field at the center of the loop that is equal to the Earth's magnetic field ($5.0 \times 10^{-4}\,T$). How much current is required?
4. A student makes a solenoid by looping a wire around a pencil. There are 600 loops, and the pencil is 12 cm long. If a current of 10 A runs through the wire, how strong is the magnetic field in the solenoid? (Ignore the Earth's magnetic field.)

FORCES DUE TO B FIELDS

A charged particle moving in a magnetic field experiences a force. The magnitude of the force is

$$F = qvB \sin \theta$$

where q is the charge, v is the speed, and θ is the angle between **v** and **B**. The direction of the force is always perpendicular to **v** and **B**. It is given by the right-hand rule, where the thumb points in the direction of **v**, the fingers point in the direction of **B**, and the **F** is normal to the palm (Figure 15.15). Note that if q is negative, we must then reverse the direction of **F**.

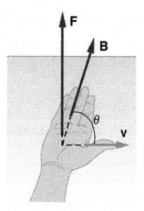

FIGURE 15.15   Right-hand rule for the force on a moving charge.

1. Find the force on the particle, where the magnetic field direction is shown in Figure 15.16.

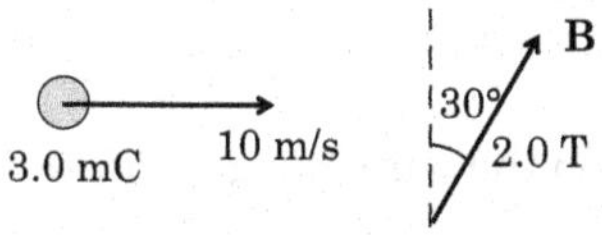

FIGURE 15.16   Particle moving in a magnetic field.

$F = (3 \times 10^{-3})(10)(2) \sin (60°) = \textbf{0.052 N}$

Direction: **out of the page**

2. Calculate the acceleration of the electron in Figure 15.17.

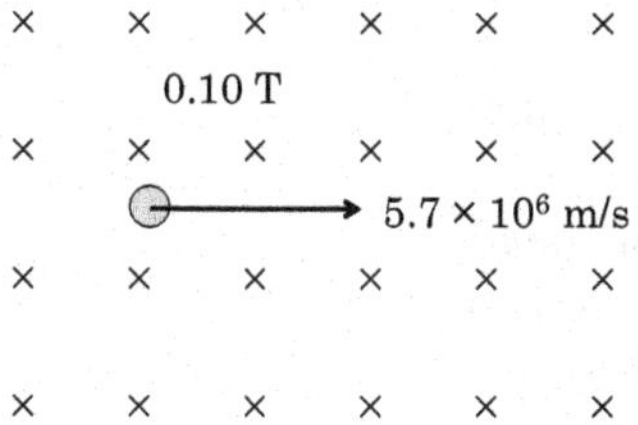

FIGURE 15.17   Electron moving in a magnetic field.

The crosses indicate that **B** points into the page.

$$F = (1.6 \times 10^{-19})(5.7 \times 10^6)(0.1) \sin (90°) = 9.12 \times 10^{-14}\,\text{N}$$

The right-hand rule gives $+y$ (up). However, because the electron is negatively charged, the force is in the $-y$ direction.

$$F_y = ma_y$$

$$-9.12 \times 10^{-14} = (9.11 \times 10^{-31})\,a_y$$

$$-1.0 \times 10^{17}\,\text{m/s}^2 = a_y$$

FORCE ON A WIRE

Current is the motion of charged particles, which experience a force in a magnetic field. The force on a straight wire is

$$F = ILB \sin \theta$$

where L is the length of the wire, and θ is the angle between **I** and **B**. We use the same right-hand rule as in the previous section, with **I** replacing **v**.

1. Calculate the force on the wire in Figure 15.18.

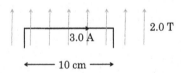

FIGURE 15.18   Wire with current in a magnetic field.

For the vertical sections, $\sin (0°) = 0$, so $F = 0$.
For the horizontal section, $\sin (90°) = 1$, so $F = ILB$.

$$F = ILB = (3)(0.1)(2) = \textbf{0.60 N}$$

Direction: **out of the page**
2. A short (2.0 cm) wire is next to a long wire (Figure 15.19). Find the force on the short wire. (The wires are attached to voltage sources, *not shown*.)

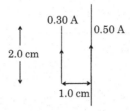

FIGURE 15.19   A 2.0-cm long wire next to a very long wire.

At the short wire, the **B** field due to the long wire points out of the page.

$B = (4\pi \times 10^{-7})(0.5)/(2\pi \times 0.01) = 1 \times 10^{-5}\,\text{T}$

$F = (0.3)(0.02)(1 \times 10^{-5}) \sin (90°) = \mathbf{6.0 \times 10^{-8}\,N}$

Direction: **+x (toward the right)**

QUIZ 15.3

1. A −7.7-mC charge travels at 4.0 m/s in the x direction. A 1.2-T magnetic field is in the x–y plane and points 17° above the x axis. What is the force on the charge?
2. Calculate the acceleration of the proton in Figure 15.20.

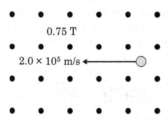

FIGURE 15.20   Proton traveling in a magnetic field.

3. A 0.30-T magnetic field points straight up. 10 A of current runs up a 15-cm-long section of wire, which is tilted 30° to the right of the **B** field. What is the force on the wire section?
4. Two long parallel wires have equal but opposite currents of 1.0 A. They are 2.0 mm apart. Consider a 10-cm section of one of the wires. What is the force on this section? Indicate whether the force is toward or away from the other wire.

CHAPTER SUMMARY

Magnetic poles	Like poles repel, opposites attract
	Outside magnet, **B** field loops from N to S
Ferromagnetic metal	Can be magnetized by external **B** field
Magnetic field due to straight wire	$B = \dfrac{\mu_0 I}{2\pi r}, \mu_0 = 4\pi \times 10^{-7}$
Magnetic field at center of one loop	$B = \dfrac{\mu_0 I}{2R}$
Magnetic field in a solenoid	$B = \mu_0 n I$
Force on particle	$F = qvB \sin \theta$
Force on wire	$F = ILB \sin \theta$

END-OF-CHAPTER QUESTIONS

1. A circular loop of wire has a 10-cm radius. It has a counterclockwise current (viewed from above) of 50 A.
   a. What is the magnetic field at the center?
   b. If a piece of iron is placed at the center, how will its magnetic poles be aligned?
2. A solenoid has 100 loops per cm, a radius of 0.50 cm, and a current of 4.0 A.
   a. What is the magnetic field in the solenoid?
   b. A bar magnet is placed next to the solenoid (Figure 15.21). Is the magnet attracted or repelled?

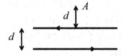

FIGURE 15.21   Solenoid next to a bar magnet.

3. Two wires are separated by $d = 2.0$ cm and have equal but opposite currents of 240 A (Figure 15.22).

FIGURE 15.22   Two wires with current running in opposite directions.

    a.   Find the magnetic field at A.

    b.   Calculate the force on a 1.0-m section of the upper wire.

4. Two parallel wires carry currents of 3.5 A (Figure 15.23). Find the magnetic field at A.

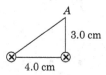

FIGURE 15.23   Two wires with current.

5. Current ($I = 30$ A) flows through a wire. An ion with charge 1.6×10^{-19} C travels toward the wire at 5.0×10^5 m/s (Figure 15.24). When the ion is $r = 0.10$ m from the wire, find the force exerted on the ion.

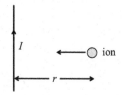

FIGURE 15.24   Ion traveling toward a current-carrying wire.

ADDITIONAL PROBLEMS

1. An electron travels at 1.0×10^6 m/s and enters a region where a 0.57-T magnetic field points into the page, and an electric field points in the $-y$ direction (Figure 15.25). Find the strength of the electric field such that the electron is not deflected up or down.

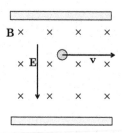

FIGURE 15.25   Electron traveling through a region with **B** and **E** fields.

2. A **B** field bends the trajectory of an electron into a circular path (Figure 15.26). Find the radius of the circle.

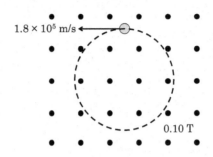

FIGURE 15.26   Electron traveling in a circular path.

3. A motor is made by sending a 0.50-A current through a square loop ($w = 5.0$ cm) in a magnetic field (0.040 T). The plane of the loop makes an angle θ with the **B** field (Figure 15.27).
   a. Calculate the magnitude of the force on the upper and lower edges of the square.
   b. Calculate the magnitude of the torque for $\theta = 20°$.

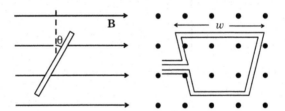

FIGURE 15.27   Square loop in a **B** field.

4. In the previous problem, calculate the force on the right edge of the square ($\theta = 20°$). Does this force produce any torque?

Electro- magnetic Induction

MAGNETIC FLUX

In this chapter, we look at *induction*, where a changing magnetic field induces a voltage in a loop. An important quantity is the *magnetic flux*, Φ, which is proportional to the number of magnetic field lines that penetrate a surface. Quantitatively, it is given by

$$\Phi = BA \cos \theta$$

where B is the magnetic field, A is the area of the surface, and θ is the angle between **B** and the surface normal (Figure 16.1).

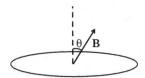

FIGURE 16.1 Magnetic field and a surface. The dashed line is perpendicular to the surface.

1. Calculate the magnetic flux through the 12-cm diameter loop (Figure 16.2). The magnetic field is 0.050 T.

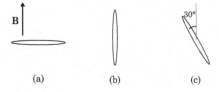

FIGURE 16.2 Loop in a **B** field. Different orientations are shown in (a), (b), and (c).

$$A = \pi(0.06^2) = 1.13 \times 10^{-2}\,\text{m}^2$$

(a) $\Phi = (0.05)(1.13 \times 10^{-2})$
   $\cos(0) = \mathbf{5.7 \times 10^{-4}\,\text{T·m}^2}$

(b) $\Phi = (0.05)(1.13 \times 10^{-2})\cos(90°) = \mathbf{0}$

(c) $\Phi = (0.05)(1.13 \times 10^{-2})$
   $\cos(60°) = \mathbf{2.8 \times 10^{-4}\,\text{T·m}^2}$

2. A square loop is in a 0.20-T magnetic field (Figure 16.3). The square collapses into a diamond. What is the change in magnetic flux?

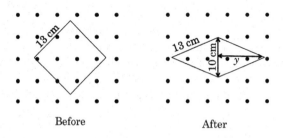

Before   After

FIGURE 16.3   Loop in a magnetic field.

The surface normal points in the same direction as **B**; $\cos(0) = 1$.

$\Phi(\text{before}) = (0.2)(0.13^2) = 3.4 \times 10^{-3}\,\text{T·m}^2$

Pythagoras: $13^2 = 5^2 + y^2 \rightarrow y = 12\,\text{cm} = 0.12\,\text{m}$

Area of the diamond: $0.10 \times 0.12 = 0.012\,\text{m}^2$

$\Phi(\text{after}) = (0.2)(0.012) = 2.4 \times 10^{-3}\,\text{T·m}^2$

$\Delta\Phi = \mathbf{-1.0 \times 10^{-3}\,\text{T·m}^2}$

LENZ'S LAW

A change in magnetic flux will induce a voltage, or electromagnetic force (emf). The emf produces current, which in turn creates a magnetic field. Lenz's law says that the direction of this induced current will produce **B** that *opposes the change in magnetic flux*. In this instance, nature abhors change and will fight against it!

1. A loop of wire is shown in Figure 16.4. A **B** field is turned on. After some time, the **B** field is turned off. What is the direction of the induced current in each case?

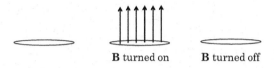

B turned on   B turned off

FIGURE 16.4   Loop of wire. A **B** field is turned on and off.

When **B** is turned on, the induced current will produce a magnetic field that points down to oppose the increase in flux. The current that does this is **clockwise** (as viewed from above).

When **B** is turned off, the induced current will produce a magnetic field that points up to oppose the decrease in flux. The current will be **counterclockwise**.

2. A bar magnet is dropped (Figure 16.5). What is the direction of the induced current in the loop?

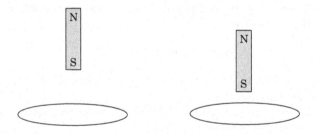

FIGURE 16.5   Bar magnet dropped above a loop of wire.

At the loop, the **B** field from the magnet points up. As the magnet drops, this **B** gets stronger. The induced current will oppose the increase.

The current will therefore run **clockwise** (viewed from above), which will produce a magnetic field that points down.

QUIZ 16.1

1. A square 4.0-cm × 4.0-cm loop is in a 0.33-T magnetic field. Two edges are perpendicular to **B**. The other two edges make an angle of 22° with **B**. Calculate the magnetic flux.

2. A loop is made with three connected wires and a movable bar (Figure 16.6). It sits in a 0.25-T magnetic field. Find the change in magnetic flux when the bar is moved to the right by 5.0 cm.

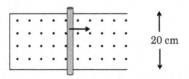

FIGURE 16.6   Moving bar in a magnetic field.

3. A **B** field points up. The normal of a metal loop is initially perpendicular to **B**. The loop is rotated so that the normal is parallel to **B**. What direction was the induced current (as viewed from above)?
4. A metal ring lies on a table to the east of a wire that runs along a north-south direction. Suddenly, current flows through the wire toward the north. What was the direction of the induced current in the ring?

FARADAY'S LAW

Faraday's law says that the magnitude of the voltage (emf) induced in a single loop is the change in flux over time,

$$\text{emf (magnitude)} = |\Delta\Phi/\Delta t|$$

In general, we will state the magnitude (absolute value) of the emf and specify the direction by using Lenz's law.

In a circuit, the emf can be modeled as a voltage source—even though there is no actual battery or other device.

1. The magnetic field in Figure 16.7 increases by 0.05 T per second. Find the emf and current.

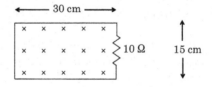

FIGURE 16.7   **B** field and a circuit.

$\Phi = BA$

Because A is constant, the change in flux is $\Delta\Phi = \Delta B{\cdot}A$

$\text{emf} = (\Delta B/\Delta t)A$

$=(0.05 \text{ T/s})(0.3)(0.15) = \mathbf{2.3 \times 10^{-3} \, V}$

$I = 2.3 \times 10^{-3}/10 = \mathbf{2.3 \times 10^{-4} \, A}$

Direction: **counterclockwise**. This produces a magnetic field pointing out of the page, which opposes the increase in flux.

2. A metal rod is free to slide on two rails (Figure 16.8). Calculate the emf while the rod is moving.

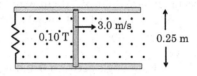

FIGURE 16.8   Moving metal rod and a magnetic field.

B is constant but the area A increases.

$\Delta\Phi = B{\cdot}\Delta A = (0.10)(0.25 \, \Delta x) = 0.025 \, \Delta x$

$\text{emf} = 0.025 \, \Delta x/\Delta t = (0.025)(3) = \mathbf{0.075 \, V}$

Direction: **clockwise**. This produces a magnetic field pointing into the page, which opposes the increase in flux.

EXAMPLES OF INDUCTION

Here are two more examples of Faraday's law.

1. A small wire loop with a radius of 6.0 cm is next to a straight wire (Figure 16.9). The current in the wire rises from 10 to 30 A in 0.020 s. Find the induced emf in the loop.

$A = \pi(0.06^2) = 0.011 \text{ m}^2$

For simplicity, take B to be the value at the loop's center.

Before: $B = (4\pi \times 10^{-7})(10)/(2\pi{\cdot}0.2) = 1.0 \times 10^{-5} \text{ T}$, out of the page

$\Phi = (1.0 \times 10^{-5})(0.011) = 1.1 \times 10^{-7}$

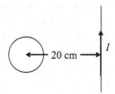

FIGURE 16.9   Loop next to a straight wire.

After: $B = (4\pi \times 10^{-7})(30)/(2\pi \cdot 0.2) = 3.0 \times 10^{-5}\,\text{T}$, out of the page

$$\Phi = (3.0 \times 10^{-5})(0.011) = 3.3 \times 10^{-7}$$

$$\text{emf} = \Delta\Phi/\Delta t = 2.2 \times 10^{-7}/0.02 = \mathbf{1.1 \times 10^{-5}\,V}$$

Direction: **Clockwise**, which will produce a **B** field into the page. This opposes the increase in magnetic flux.

2. The plane of a rectangular 3.0-cm × 5.0-cm wire loop is initially 22° to a 1.1-T magnetic field. It rotates to 50° in 5.0×10^{-2} s. What was the average emf during the rotation?

$\Phi = BA \cos \theta$, where θ is the angle between **B** and the normal to the loop.

$$A = (0.03)(0.05) = 1.5 \times 10^{-3}\,\text{m}^2$$

Before: $\Phi = (1.1)(1.5 \times 10^{-3}) \cos (68°) = 6.2 \times 10^{-4}$

After: $\Phi = (1.1)(1.5 \times 10^{-3}) \cos (40°) = 1.26 \times 10^{-3}$

$$\text{emf} = \Delta\Phi/\Delta t = 6.4 \times 10^{-4}/5.0 \times 10^{-2} = \mathbf{1.3 \times 10^{-2}\,V}$$

QUIZ 16.2

1. A square 1.5-m × 1.5-m loop is made of metal wire. A magnetic field is perpendicular to the plane of the loop and points out of the page. The field decreases at a rate of -0.20 T/s. Calculate the emf.
2. A loop is defined by two metal rails and two metal rods (Figure 16.10). Each rod moves at 8.0 m/s. Calculate the emf.

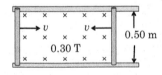

FIGURE 16.10   Two rods moving toward each other in a magnetic field.

3. A 5.0-mm radius metal ring lies on a table, 40 cm to the east of a wire that runs along a north-south direction. Current begins to run through the wire toward the north. It takes 0.010 s to reach 20 A. What is the induced emf in the ring?
4. A 20-cm diameter loop is rotated in a 0.50-T magnetic field. What is the average emf induced, given that the plane of the loop is initially perpendicular to the field and is rotated to be parallel to the field in 5.0 ms?

GENERATORS

The previous examples have involved a single loop of wire. A coil with N loops, or turns, will have a larger emf than a single loop:

$$\text{emf (magnitude)} = N|\Delta\Phi/\Delta t|$$

A *generator* can be made from a coil that rotates in a magnetic field (Figure 16.11). The emf induced in the coil is sinusoidal (AC). The peak emf is

$$\text{Peak emf} = NAB\omega$$

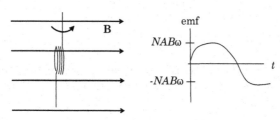

FIGURE 16.11   Coil rotating in a **B** field.

where A is the loop area, B is the magnetic field, and ω is the rotational speed (rad/s).

1. A 1200-turn, 40-cm radius coil is in the Earth's magnetic field (5.0×10^{-4} T). It takes 10 ms for the plane of the coil to rotate from parallel to perpendicular to the field. What is the average emf induced in the coil?

   $A = \pi(0.4^2) = 0.50 \text{ m}^2$

   Before: $\Phi = (5.0 \times 10^{-4})(0.5) \cos(90°) = 0$

   After: $\Phi = (5.0 \times 10^{-4})(0.5) \cos(0) = 2.5 \times 10^{-4}$

   $\text{emf} = 1200(2.5 \times 10^{-4}/0.01) = \textbf{30 V}$

2. Calculate the peak voltage of a generator that rotates its 50-turn, 0.10-m diameter coil at 60 rpm in a 0.80-T field.

   $A = \pi(0.05^2) = 7.9 \times 10^{-3} \text{ m}^2$

   $\omega = (60 \text{ rotations/min})(1 \text{ min}/60 \text{ s})(2\pi \text{ rad/rotation}) = 6.28 \text{ rad/s}$

   $\text{Peak emf} = (50)(7.9 \times 10^{-3})(0.8)(6.28) = \textbf{2.0 V}$

TRANSFORMERS

A transformer consists of two coils wrapped around a ferromagnetic metal (Figure 16.12). The primary coil has an alternating current, which produces a sinusoidal **B** field. This field induces an emf in the secondary coil. The magnetic flux passing through each coil is the same.

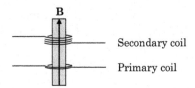

FIGURE 16.12   A transformer.

The emf in the primary coil is $V_p = N_p\Delta\Phi/\Delta t$, where N_p is the number of turns. The emf in the secondary coil is $V_s = N_s\Delta\Phi/\Delta t$. Dividing the second equation by the first yields

$$V_s/V_p = N_s/N_p$$

If we want the voltage in the secondary coil to be large, then we make the number of turns in the secondary coil larger than that of the primary coil. Note that power $(P = IV)$ remains constant; if V is increased, then I must decrease by the same factor.

1. A transformer converts 240 V to 120 V. What is the ratio of primary to secondary loops?

$$V_s/V_p = N_s/N_p$$

$$120/240 = N_s/N_p$$

$$0.5 = N_s/N_p$$

$$\mathbf{2.0} = N_p/N_s$$

2. In the previous problem, a single device uses the 120 V; 0.60 A flow through the device. How much current flows through the primary coil?

The secondary coil provides power $P = IV = (0.6)(120) = 72$ W. The primary coil provides the same power.

$$72 = I\,(240)$$

$$\mathbf{0.30\ A} = I$$

QUIZ 16.3

1. An emf is induced by rotating a 50-turn, 20-cm diameter coil in a 0.32-T magnetic field. What average emf is induced, given the plane of the coil is originally perpendicular to the magnetic field and is rotated to be parallel to the field in 0.050 s?
2. Calculate the peak voltage of a generator that rotates its 100-turn, 0.10-m diameter coil at 6,000 rpm in a 3.0-T field.
3. A plug-in transformer supplies 9.0 V to a video game system. How many turns are in its secondary coil if its input voltage is 120 V and the primary coil has 600 turns?
4. In the previous question, suppose the video game takes 1.2 A of current. How much current goes through the primary coil?

CHAPTER SUMMARY

Magnetic flux	$\Phi = BA \cos \theta$
Lenz's law	Induced current opposes change in Φ
Faraday's law	emf (magnitude) $= N\lvert\Delta\Phi/\Delta t\rvert$
	N = number of turns in coil
Generator	Peak emf $= NAB\omega$
Transformer	$V_s/V_p = N_s/N_p$

END-OF-CHAPTER QUESTIONS

1. A bar magnet sits on a table with N pointing up. A loop is above the magnet and parallel to the table. The loop is dropped. What is the direction of the induced current (seen from above)?
2. A wire loop has the shape of an equilateral triangle with 12-cm sides (Figure 16.13). Its plane is initially parallel to a 0.16-T magnetic field. It is rotated to be perpendicular to the field in 2.0 ms.

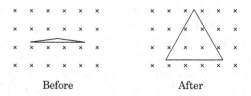

Before   After

FIGURE 16.13   Triangular loop in a magnetic field.

    a. What average emf is induced?
    b. What is the direction of the induced current?

3. A circuit contains a movable metal rod and a resistor (Figure 16.14). The circuit is in a 0.30-T magnetic field.

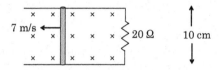

FIGURE 16.14 Moving metal rod and a magnetic field.

   a. Find the induced current in the circuit.
   b. Calculate the force on the bar due to the 0.30-T magnetic field.
4. A 200-turn, 7.0-cm diameter coil rotates at 300 rpm in a 0.040-T magnetic field. The coil is attached to a 500-Ω load. Calculate the current flowing through the load.
5. A transformer is made of a 20-turn primary coil and 100-turn secondary coil. A 120-V alternating-current power supply sends 5.0 A through the primary coil. What is the voltage and current provided by the secondary coil?

ADDITIONAL PROBLEMS

1. A 0.32-m diameter loop of wire sits on a table. The plane of the circular loop is perpendicular to a 0.25-T magnetic field. Someone changes the loop's shape into a square. Assuming it takes 0.10 s for this change in shape to happen, what was the average induced emf?
2. A small ring (2.0-cm diameter) and large loop (1.4-m diameter) are concentric and sit on a table. The small ring has a resistance of 0.20 Ω. The current in the large loop increases from 0 to 30 A in 0.015 s. Viewed from above, the current is clockwise. Calculate induced current in the small ring.
3. A solenoid is oriented vertically. A metal ring sits on top of the solenoid and is concentric with the solenoid. Current begins to flow through the solenoid such that **B** points up. During the current ramp up, find:
   a. The direction of the induced current in the ring (as viewed from above).
   b. The direction of the force on the ring (due to the solenoid).
4. If a solid piece of metal is in a region where the **B** field changes, it will develop induced currents that oppose the change in magnetic flux.

These are called *Eddy currents*. We can simulate this effect by considering a rectangular loop of wire that moves into a region of magnetic field (Figure 16.15).

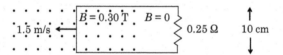

FIGURE 16.15   Circuit moving into a region where $B = 0.30$ T.

   a.   Calculate the induced current.
   b.   Calculate the force on the rectangular loop (due to the magnetic field).

Electromagnetic Waves

17

WAVE PROPERTIES

Consider an antenna where current runs up and down at a high frequency (Figure 17.1). At a snapshot in time, there is a dipole (i.e., positive charge at one end and negative charge at the other end). This dipole results in an **E** field outside the antenna. In the picture, the **B** field is perpendicular to the page. The **E** and **B** fields propagate away from the antenna. This is an *electromagnetic wave*.

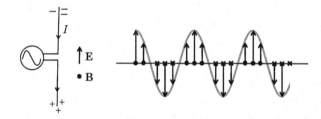

Antenna   Electromagnetic wave

FIGURE 17.1 Antenna with a current that runs up and down. An electromagnetic wave travels away from the antenna.

An electromagnetic wave in vacuum or air has a speed $c = 3.0 \times 10^8$ m/s. The wave speed equation (Chapter 9) is

$$c = \lambda f$$

where λ is the wavelength (m), and f is the frequency (s^{-1} or Hz). The human eye can see wavelengths from 400 to 700 nm, a range called "visible light."

1. A radio station broadcasts at 1250 kHz. What is the wavelength?

$3.0 \times 10^8 = \lambda \, (1.25 \times 10^6)$

240 m = λ

2. What frequencies of light can be seen by the human eye?

$3.0 \times 10^8 = (4 \times 10^{-7})f, \quad 3.0 \times 10^8 = (7 \times 10^{-7})f$

$7.5 \times 10^{14} \, \text{Hz} = f, \qquad 4.3 \times 10^{14} \, \text{Hz} = f$

The eye can see **4.3×10^{14} to 7.5×10^{14} Hz**

POLARIZATION

In an electromagnetic plane wave, **E** and **B** are perpendicular to each other (Figure 17.2). The *polarization* refers to the direction of **E**.

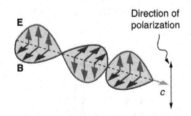

FIGURE 17.2   Electromagnetic wave.

A *polarizing filter* lets through the component of **E** that is parallel to the polarization axis and blocks the perpendicular component. The electric-field amplitude passing through the filter is

$$E = E_0 \cos\theta$$

where E_0 is the incident amplitude (the electric field before the filter), and θ is the angle between **E** and the polarization axis.

The *intensity* (W/m²) of light is proportional to E^2. The light intensity I that passes through a polarizing filter is

$$I = I_0 \cos^2\theta$$

where I_0 is the incident intensity.

Sources such as light bulbs emit light that is randomly polarized, or *unpolarized*. When passed through a polarizing filter, the light becomes polarized, with $E = E_0/\sqrt{2}$ and $I = I_0/2$.

1. Unpolarized light travels through two polarizing filters with perpendicular axes (Figure 17.3). What fraction of the intensity passes through each filter?

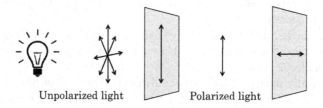

Unpolarized light   Polarized light

FIGURE 17.3 Unpolarized light travels through a polarizing filter.

After the first filter, $I = I_0/2$, so the fraction is **0.5**. The light is then vertically polarized.
The vertically polarized light is 90° to the second filter's axis.
After the second filter, $I = I_0\cos^2(90°) = $ **0**. No light makes it through.

2. Vertically polarized light travels through two polarizing filters (Figure 17.4). The first filter has an axis 22° from vertical. The second filter has a horizontal axis. What fraction of the light intensity passes through both filters?

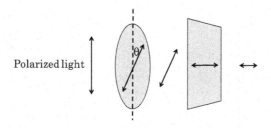

Polarized light

FIGURE 17.4 Polarized light travels through two filters.

Let I_0 be the intensity of the incoming polarized light.
Light passing through the first filter is polarized 22° from vertical. Its intensity is

$$I = I_0\cos^2(22°) = I_0(0.93^2) = I_0(0.86)$$

Light passing through the second filter has an intensity

$$I = I_0(0.86) \cos^2(68°) = I_0(0.12),$$ so the fraction is **0.12**.

QUIZ 17.1

1. Extremely-low-frequency radio waves of 1.0 kHz are used to communicate with submarines. Find the wavelength.
2. A laser emits ultraviolet (UV) light with wavelength 325 nm. What is the frequency?
3. Unpolarized light (0.060 W/cm²) passes through two polarizing filters with axes 39° to each other. How much intensity makes it through both filters?
4. Vertically polarized light passes through a polarizing filter with an axis 18° to horizontal. What fraction of the electric-field strength and intensity transmits through the filter?

TWO-SLIT INTERFERENCE

As we saw in Chapter 9, when two waves overlap, their amplitudes add. We will now look at light passing through two narrow slits (Figure 17.5). Light passing through a narrow slit propagates outward in a semicircular fashion. At the screen, the two light waves add together. If the path-length difference is a multiple of the wavelength (0, λ, 2λ, ...), we have constructive interference (bright fringe, or maximum). If the difference is $\frac{1}{2}\lambda$, $1\frac{1}{2}\lambda$, $2\frac{1}{2}\lambda$, ..., we have destructive interference (dark fringe, or minimum).

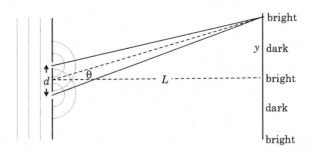

FIGURE 17.5   Light passing through two narrow slits, resulting in bright and dark fringes on a screen.

Let d be the distance between the slit centers. Because $L \gg d$, the two rays (diagonal lines in the figure) are almost parallel and make an angle θ with horizontal. The interference conditions are

- Constructive interference: $d \sin \theta = m\lambda$
- Destructive interference: $d \sin \theta = (m + \frac{1}{2})\lambda$

where m is an integer ($0, \pm1, \pm2$) and is called the *order* of the interference. The bright $m = 0$ fringe occurs at $\theta = 0$ and is called the *central maximum*.

1. A red (650 nm) laser shines through two slits 0.10 mm apart. A screen is 2.0 m away. What is the distance between the bright fringes?

   The $m = 0$ bright fringe occurs at $y = 0$ (central maximum).

   $m = 1$: $(10^{-4}) \sin \theta = 6.5 \times 10^{-7}$

   $\sin \theta = 6.5 \times 10^{-3}$

   $\theta = \sin^{-1}(6.5 \times 10^{-3}) = 0.37°$

   $\tan \theta = y/L$

   $6.5 \times 10^{-3} = y/2.0$ (notice how $\sin \theta \approx \tan \theta$ for small θ)

   $\mathbf{1.3 \times 10^{-2}\,m = y}$

2. The constructive interference condition for multiple slits is the same as for two slits. A *diffraction grating* has lines that reflect light; these lines act as slits. Suppose a diffraction grating has 1000 lines per cm. At what angle will the second-order maximum be for 532-nm green light?

   $d = 1/(1000 \text{ lines/cm}) = 0.001 \text{ cm} = 10^{-5}\,m$

   $d \sin \theta = 2\lambda$ ($m = 2$)

   $(10^{-5}) \sin \theta = 2(5.32 \times 10^{-7})$

   $\sin \theta = 0.106$

   $\theta = \sin^{-1}(0.106) = \mathbf{6.1°}$

SINGLE SLIT AND CIRCULAR APERTURE

In the previous section, we assumed each slit was very narrow. Light waves emanated from the slits in a semicircular pattern. We now consider a slit of width, D (Figure 17.6). Light rays coming from different parts of the slit will interfere with each other. This results in a central maximum with fringes on either side. The minima (dark fringes) occur at

$$D \sin \theta = m\lambda$$

where $m = \pm1, \pm2, \ldots$

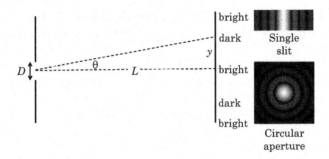

FIGURE 17.6   Light passing through a narrow opening (slit or circular aperture).

For a circular aperture (pinhole) of diameter, D, we get a central bright spot with rings. The first dark ring (minimum) occurs at

$$D \sin \theta = 1.22\lambda$$

1. Blue light (450 nm) passes through a 0.50-mm wide slit and hits a screen 1.2 m away. What is the width of the central maximum?

   $m = 1$: $(5 \times 10^{-4}) \sin \theta = 4.5 \times 10^{-7}$

   $\sin \theta = 9 \times 10^{-4}$

   $\tan \theta = y/L$

   $9 \times 10^{-4} = y/1.2$ ($\sin \theta \approx \tan \theta$)

   $1.1 \times 10^{-3} = y$

   The $m = \pm 1$ dark fringes occur at $\pm 1.1 \times 10^{-3}$ m
   The width of the central maximum is $2y = \mathbf{2.2 \times 10^{-3}\,m}$

2. A red laser (650 nm) pointer shoots a 1.0-mm diameter beam at a target 1.0 km away. What is the diameter of the central maximum at the target?

   The diameter of the aperture is $D = 1.0$ mm.

   $(1 \times 10^{-3}) \sin \theta = 1.22 (6.5 \times 10^{-7})$

   $\sin \theta = 7.9 \times 10^{-4}$

   $\tan \theta = y/L$

$7.9 \times 10^{-4} = y/1000$ $\qquad$ $(\sin \theta \approx \tan \theta)$

$0.79 = y$

The $m = 1$ dark ring has a radius of 0.79 m.
The diameter of the central maximum is $2y = \textbf{1.6 m}$

QUIZ 17.2

1. Light of wavelength 580 nm passes through two slits 0.29 mm apart. A screen is 50 cm from the slits. What is the distance between the central maximum and the first dark fringe?
2. Red light (wavelength 660 nm) travels through three vertical slits. The distance between the centers of adjacent slits is 0.10 mm. What is the angle for the third-order maximum?
3. A green laser (532 nm) passes through a 0.040-mm wide slit and hits a piece of paper 30 cm away. What is the width of the central maximum on the paper?
4. In the previous problem, the slit is replaced by a 0.065-mm diameter pinhole. What is the diameter of the central maximum on the paper?

REFLECTION AND REFRACTION

A *ray* is an arrow that represents the direction that a wave moves, or propagates. Suppose light strikes a smooth surface. The *angle of incidence*, θ_1, is the angle that the ray makes with the surface normal (Figure 17.7). The *angle of reflection* equals the angle of incidence.

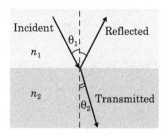

FIGURE 17.7   Light incident on a surface.

Light can transmit through a transparent medium but is slowed down. The speed of light in the medium is

$$v = c/n$$

where c is the speed of light in vacuum and n is the *index of refraction*. Typical media are air ($n = 1$), water ($n = 1.33$), and glass ($n = 1.5$). When light travels from a medium with index n_1 to one with n_2, its direction changes, or *refracts*. The angles of the incident and refracted rays are related by *Snell's law*:

$$n_1 \sin \theta_1 = n_2 \sin \theta_2$$

1. A glass aquarium is filled with water. A laser beam travels through the water and hits the glass bottom with an incident angle of 27°. What is the angle of the light that travels into the glass?

$n_1 \sin \theta_1 = n_2 \sin \theta_2$

$(1.33) \sin (27°) = (1.5) \sin \theta_2$

$0.40 = \sin \theta_2$

$\theta_2 = \sin^{-1} (0.40) = \mathbf{24°}$

2. A fish is 1.0 m under water (Figure 17.8). Sunlight reflecting off the fish transmits into the air. To an observer in the air, the fish appears to be at a depth, d. Find d.

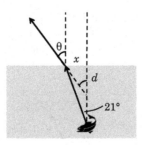

FIGURE 17.8   A fish under water. Arrows indicate a light ray coming from the fish.

Find x: $\tan (21°) = x/1.0$

$0.38 = x$

Find θ: $(1.33) \sin (21°) = (1) \sin \theta$

$0.48 = \sin \theta$

$\theta = \sin^{-1} (0.48) = 28°$

Find d: $\tan(28°) = x/d$

$0.53 = 0.38/d$

$0.53d = 0.38$

$d = \textbf{0.72 m}$

THIN FILMS

A thin film (e.g., layer of oil or thin glass slide) has two surfaces. Light reflected from the bottom surface interferes with the light reflected from the top (Figure 17.9). This yields constructive interference for certain wavelengths and can give the film color.

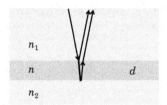

FIGURE 17.9 Film of thickness, d, and index, n. (Some light goes into n_2, not shown.)

To determine the interference conditions, we note the following:

- The wavelength in a medium is λ/n, where λ is the wavelength in air. The wavelength is shortened because the medium slows down the light wave.
- If the wave encounters a higher n medium, the reflection undergoes a *phase shift* where crests become troughs.

Consider light reflected from a thin film at near-normal incidence ($\theta \approx 0$). The interference conditions are shown in the table, where d is the film thickness and $m = 0, 1, 2, \ldots$

	0 or 2 Rays Have Phase Shift	1 Ray Has Phase Shift
Constructive	$2d = m\lambda/n$	$2d = (m + \frac{1}{2})\lambda/n$
Destructive	$2d = (m + \frac{1}{2})\lambda/n$	$2d = m\lambda/n$

1. Light reflects from a soap bubble. What are the two smallest thicknesses that will give constructive interference for red light (650 nm)?

   The soap bubble is essentially a water film ($n = 1.33$) in air ($n_1 = n_2 = 1$).

The ray reflecting off the top water surface will undergo a phase shift.
The ray reflecting off the bottom water surface will not.
Constructive interference:

$$2d = (m + \tfrac{1}{2})\lambda/n$$

$$2d = (m + \tfrac{1}{2})(650 \text{ nm})/1.33 = 244 \text{ nm}, 733 \text{ nm}$$

$$d = \textbf{122 nm, 367 nm}$$

2. A 400-nm thin layer of oil ($n = 1.4$) is on a glass slide. What wavelength of reflected visible light will experience constructive interference?

$n_1 = 1$ (air is assumed) and $n_2 = 1.5$
The ray reflecting off the top oil surface will undergo a phase shift.
The ray reflecting off the bottom oil surface will undergo a phase shift.
Constructive interference:

$$2d = m\lambda/n$$

$$2(400 \text{ nm}) = m\lambda/1.4$$

$$(1120 \text{ nm})/m = \lambda$$

$$\lambda = 1120 \text{ nm}, 560 \text{ nm}, 373 \text{ nm}, \ldots$$

The visible wavelength is **560 nm**

QUIZ 17.3

1. Glass has an index of refraction that depends on wavelength, a property called *dispersion*. For blue light (470 nm), $n = 1.524$. For red light (660 nm), $n = 1.512$. Find the angle θ for blue and red light (Figure 17.10).

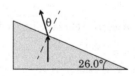

FIGURE 17.10 Light traveling through a glass prism and transmitting into the air.

2. A person stands at the edge of a pool and holds a laser pen 1.0 m above the water surface. The laser beam hits the water surface 1.5 m from the edge. If the water is 2.0 m deep, find the location of the laser spot at the bottom of the pool (distance from the edge).
3. Two glass microscope slides are separated by an air gap. Find the smallest air-gap thickness such that 500-nm reflected light has constructive interference.
4. Sometimes manufacturers deposit a thin "antireflective" coating on glass to minimize the intensity of reflected light. For a coating with $n = 1.36$, what is the minimum thickness that will give destructive interference for 544-nm light?

CHAPTER SUMMARY

Light in vacuum	$c = \lambda f$
	$c = 3.0 \times 10^8$ m/s
Light through a polarizing filter	Polarized light: $I = I_0 \cos^2\theta$
	Unpolarized light: $I = I_0/2$
	I (W/m²) is proportional to E^2
Two-slit interference	Maxima: $d \sin\theta = m\lambda$
	Minima: $d \sin\theta = (m + \frac{1}{2})\lambda$
Single slit	Minima: $D \sin\theta = m\lambda$
Circular aperture	First minimum: $D \sin\theta = 1.22\lambda$
Index of refraction (n)	$v = c/n$
Snell's law	$n_1 \sin\theta_1 = n_2 \sin\theta_2$
Thin-film interference	If 1 ray has a phase shift,
	Constructive: $2d = (m + \frac{1}{2})\lambda/n$
	Destructive: $2d = m\lambda/n$
	If 0 or 2 rays have a phase shift,
	Constructive: $2d = m\lambda/n$
	Destructive: $2d = (m + \frac{1}{2})\lambda/n$

END-OF-CHAPTER QUESTIONS

1. Unpolarized ultraviolet (UV) light of wavelength 300 nm travels through two polarizing filters. The angle between the filter axes is 67°.
   a. What is the frequency of light?
   b. The light intensity incident on the first filter is 3.0×10^{-3} W/cm². What is the intensity that transmits through both filters?

2. Red light (wavelength 660 nm) travels through three vertical slits. The distance between the centers of adjacent slits is 0.10 mm. Each slit has a width of 0.030 mm. The light hits a screen 3.0 m away.
   a. What is the angle (degrees) for the first-order maximum?
   b. The center slit is blocked. What is the distance between fringes on the screen?
   c. Two slits are blocked. What is the width of the central maximum on the screen?

3. Sunlight illuminates a piece of paper, which has a 0.45-mm diameter pinhole. A piece of cardboard is 4.0 m away and parallel to the paper. What is the diameter of the spot on the cardboard? Approximate the average wavelength of sunlight as 500 nm.

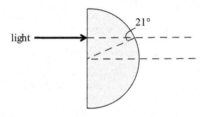

FIGURE 17.11   Light ray normal to a plastic hemisphere.

4. A light ray is normal to the flat surface of a plastic hemisphere (Figure 17.11). The index of refraction of the plastic is 1.43. The hemisphere is surrounded by air.
   a. What is the speed of light in the plastic?
   b. At what angle (with respect to horizontal) does the light ray emerge from the hemisphere?

5. A thin (1.0 μm) film of water is between two glass microscope slides. Which reflected light wavelengths between 450 and 800 nm will have constructive interference?

ADDITIONAL PROBLEMS

1. Light (wavelength 500 nm) from a lightbulb passes through two polarizing filters and a narrow slit (width 0.20 mm). The light then hits a screen, which is 1.7 m away from the slit (Figure 17.12). Initially, the axes of the polarizing filters are parallel ($\theta = 0$), and the light intensity passing through the slit is 0.040 W/cm^2.
   a. What is the full width of the central maximum on the screen?
   b. The second polarizing filter is rotated to $\theta = 37°$. What is the light intensity passing through the slit?

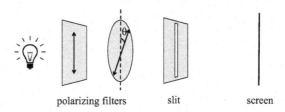

polarizing filters   slit   screen

FIGURE 17.12 Unpolarized light passing though filters and a slit.

2. Two glass slides are separated by an air gap. The air gap is a wedge with an angle of 1.0°. A 550-nm light shines on the slides at near-normal incidence. What is the distance between reflected fringes?
3. Light inside glass travels toward the glass-air interface. For what incident angle will the refracted angle be 90°? (This is called *total internal reflection* because no light propagates into the air).
4. A 290-nm thin film of an unknown substance is on a glass slide. A spectrometer measures strong reflections for 375- and 750-nm light (near-normal incidence). What is the index of refraction of the thin film?

Optics

LENSES

A lens is typically made of glass or plastic and has one or two curved surfaces that refract light. The curvature is designed so that parallel light rays all converge at a *focal point* (*f*); see Figure 18.1. Rays that emanate from *f* will be made parallel. A light ray that passes through the center of a thin lens does not change its direction.

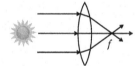

FIGURE 18.1 Left: A lens focuses parallel light onto a spot at *f*. Right: Light rays that emanate from *f* are made parallel by the lens.

Ray tracing uses these rules to find the image produced by a lens. In the picture, an *object* is to the left of the focal point, *f* (Figure 18.2). We draw three rays: one horizontal, one passing through *f*, and one passing through the center of the lens. These rays converge on a point to the right of the lens. This is called the *image*.

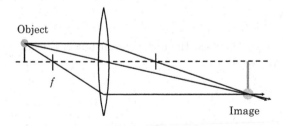

FIGURE 18.2 Ray-tracing diagram.

A person to the right of the image would see it as being larger than the object and inverted (upside down).

1. An object is a distance 2f to the left of a lens (Figure 18.3). Where is the image?

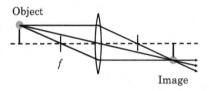

Object

f

Image

FIGURE 18.3 Ray-tracing diagram for an object a distance 2f from the lens.

In this case, the rays are symmetric, so the image is 2f to the right of the lens. This is called a *real image*. Someone could get very close to the image, as if it were "real."

2. An object is 0.5f to the left of a lens (Figure 18.4). Where is the image?

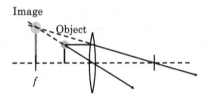

Image

Object

f

FIGURE 18.4 Ray-tracing diagram for an object 0.5f from the lens.

In this case, the rays extrapolate to a point on the same side as the object. The image is a distance, f, to the left of the lens. This is called a *virtual image*. It appears as if it is twice as tall as the object but further away.

THE LENS EQUATION

An object has a distance, s, from the lens. The image is a distance, s', from the lens, calculated from the lens equation:

$$\frac{1}{s} + \frac{1}{s'} = \frac{1}{f}$$

If $s' > 0$, it is a real image and is on the opposite side from the object. If $s' < 0$, it is a virtual image and is on the same side as the object.

The *magnification*

$$m = -s'/s$$

gives the height of the image as compared to the object. If $m > 0$, the image is upright. If $m < 0$, the image is inverted.

1. An object is 30 mm from a 25-mm focal length lens. Find the location of the image, whether it is real or virtual, the magnification, and whether it is upright or inverted.

$$\frac{1}{30} + \frac{1}{s'} = \frac{1}{25}$$

$1/s' = 1/25 - 1/30 = 6.67 \times 10^{-3}$

$s' = $ **150 mm, opposite side from object, real image**

$m = -150/30 = $ **−5.0, inverted**

2. A person holds a 10-cm magnifying lens 2.5 cm from an insect. Find the location of the image, whether it is real or virtual, the magnification, and whether it is upright or inverted.

$$\frac{1}{2.5} + \frac{1}{s'} = \frac{1}{10}$$

$1/s' = 1/10 - 1/2.5 = -0.30$

$s' = $ **−3.3 cm, same side as object, virtual image**

$m = 3.3/2.5 = $ **1.3, upright**

QUIZ 18.1

1. An object is placed 25 cm to the left of a 10-cm focal length lens. Draw a ray-tracing diagram that shows the image.
2. For the previous problem, calculate the position and magnification of the image. Is the image real or virtual?
3. In a large lecture hall, the professor puts a light bulb 31 cm from a 30-cm focal length lens. She places a screen on the other side, which

shows an in-focus image of the bulb. Find the distance from the lens to the screen and the magnification.

4. A dermatologist looks through a lens ($f = 5.0$ cm) to examine a 3.2-mm diameter mole. The mole is 4.0 cm from the lens. From the dermatologist's perspective, what is the location and size of the image? Indicate whether the image is upright or inverted.

MAGNIFIERS

A tall mountain far away looks small, but a child up close looks tall. The *apparent size* is given by the angle θ_0 subtended by the object (Figure 18.5). The closer your eye gets to the object, the larger is the object's apparent size.

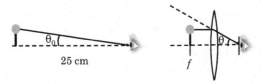

25 cm

FIGURE 18.5   The apparent size of an object is determined by θ.

The human eye is good at seeing distant objects. However, if an object is closer than 25 cm, it gets blurry. A *magnifier* produces a virtual image that is far away and therefore easy to look at. When the object is in the focal plane (i.e., a distance, f, from the lens), the virtual image is infinitely far away. The *angular magnification* is the ratio between the apparent size of the image and the apparent size seen by an unaided eye 25 cm away,

$$M = \frac{\theta}{\theta_0} = \frac{25\ \text{cm}}{f}$$

1. A magnifier has a focal length of 100 mm. An object is 98 mm from the magnifier, very close to the focal plane. Find the location of the virtual image.

$$\frac{1}{98} + \frac{1}{s'} = \frac{1}{100}$$

$1/s' = 1/100 - 1/98 = -2.04 \times 10^{-4}$

$s' = -4900$ mm, same side as object
(We can consider this to be infinitely far away.)

2. A person looks through a magnifier ($f = 5.0$ cm) at an insect that is 3.5 mm long. What is the angle subtended by the image?

$M = 25$ cm/5 cm $= 5$

When looking at the insect without a magnifier,

$\tan \theta_0 = 0.35$ cm/25 cm $= 0.014$

$\theta_0 = \tan^{-1}(0.014) = 0.80°$

$\theta = 0.80° \times M = $ **4.0°**

MICROSCOPES

A compound microscope consists of two lenses, an objective and eyepiece. The object is slightly below the focal plane of the objective. The objective produces a real image of the object. (If it is a 10× objective, for example, then $m = -10$.) The eyepiece acts as a magnifier, with the real image as its "object." The total magnification is $m \times M$.

1. A microscope (Figure 18.6) has an objective ($f_o = 16$ mm) and eyepiece ($f_e = 50$ mm). What is the position of the specimen?

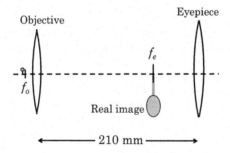

FIGURE 18.6   A microscope. The objective produces a real image at f_e.

The real image is at $s' = 210 - 50 = 160$ mm

$$\frac{1}{s} + \frac{1}{160} = \frac{1}{16}$$

$1/s = 1/16 - 1/160 = 5.6 \times 10^{-2}$

$s = $ **18 mm**

2. In the previous problem, find the total magnification of the microscope.

$m = -160/18 = -9$

$M = 25$ cm/5 cm $= 5$

Total magnification $= -9 \times 5 = -45$

QUIZ 18.2

1. A magnifier is useful when the object distance equals the focal length. Suppose a magnifier with a focal length of 10 cm is held 2.0 cm from an object. Find the position and magnification of the virtual image.
2. Fred looks at his 8.0-cm long finger using a magnifier ($f = 12.5$ cm). What angle is subtended by the image?
3. In a microscope, the real image is 165 mm from the objective. Find the magnification of an objective lens that has a focal length of 3.24 mm.
4. In the previous problem, find the total magnification if the eyepiece has a focal length of 10 cm.

TELESCOPES

A telescope magnifies distant objects. An object that is far away from the objective subtends a small angle θ_o (Figure 18.7). The objective produces a real image at f_o. This real image is the "object" for the eyepiece. The angular magnification is

$$M = \frac{\theta_e}{\theta_o} = -\frac{f_o}{f_e}$$

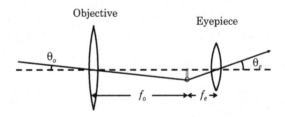

FIGURE 18.7 A telescope. The real image is a distance, f_o, from the objective and f_e, from the eyepiece.

1. An object is 50 m away from a telescope with an objective focal length of 60 cm. Where is the real image?

$$\frac{1}{50} + \frac{1}{s'} = \frac{1}{0.60}$$

$1/s' = 1/0.6 - 1/50 = 1.647$

$s' = \mathbf{0.61\ m}$ (very close to the objective focal plane)

2. A 12-m tall tree is 800 m away. You look at the tree through a telescope with objective and eyepiece focal lengths of 150 and 50 mm. What angle is subtended by the image?

$M = -150/50 = -3$

$\tan \theta_o = 12/800 = 0.015$

$\theta_o = \tan^{-1}(0.015) = 0.86°$

$\theta_e = -3 \times 0.86 = \mathbf{-2.6°}$ (the image is inverted)

RESOLUTION

In Chapter 17, we saw that light passing through a circular aperture has a central maximum, or spot. If we have two point sources, they will produce two spots on the screen (Figure 18.8).

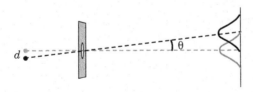

FIGURE 18.8   Light from two point sources passing through a circular aperture.

The *Rayleigh criterion* states that two sources can be resolved when the maximum of one spot lies on the 0 of the other. This occurs when

$$D \sin \theta = 1.22\ \lambda$$

where θ is the angle subtended by the two points, D is the aperture diameter, and λ is the light wavelength. Angles greater than θ will be easily resolved, and smaller angles result in an unresolved blob. Even though this is illustrated for an aperture and screen, the equation is reasonably accurate for a microscope or telescope with a circular aperture.

1. The Hubble Space Telescope aperture has a diameter of 2.4 m. What is the angle between two just-resolvable stars? Assume an average light wavelength of 550 nm.

$$2.4 \sin \theta = 1.22 \, (5.5 \times 10^{-7} \, \text{m})$$

$$\sin \theta = 2.8 \times 10^{-7}$$

$$\theta = \sin^{-1} (2.8 \times 10^{-7}) = \mathbf{1.6 \times 10^{-5} \, degrees}$$

2. A microscope aperture is 8.0 mm in diameter. A specimen is 14 mm below the aperture. How close can two cells be to each other and still be resolved?

We'll again assume $\lambda = 550$ nm.

$$8 \sin \theta = 1.22 \, (5.5 \times 10^{-7} \, \text{m})$$

$$\sin \theta = 8.4 \times 10^{-8}$$

$$\tan \theta = d/14$$

$$8.4 \times 10^{-8} = d/14 \qquad (\sin \theta \approx \tan \theta)$$

$$\mathbf{1.2 \times 10^{-6} \, m = d}$$

QUIZ 18.3

1. If an object is too close to a telescope, the real image will not be in the right place. Suppose someone uses a telescope with a 10-cm focal length objective to look at an object 25 cm away. Where will the real image be?
2. A biologist uses a telescope with objective and focal lengths of 200 mm and 50 mm to view a giraffe 450 m away. The image of the giraffe subtends an angle of 2.5°. How tall is the giraffe?
3. A spy satellite is 300 km above the earth. It aims its telescope straight down. What aperture diameter would be required to resolve two points 20 cm apart? Assume a wavelength of 550 nm.

4. Two dyed mitochondria emit 600 nm light. They are 14 mm below the microscope objective (aperture diameter 7.0 mm). How close can the mitochondria be to each other and still be resolved?

CHAPTER SUMMARY

Lens equation	$\dfrac{1}{s} + \dfrac{1}{s'} = \dfrac{1}{f}$
Magnification	$m = -s'/s$
Angular magnification of a magnifier	$M = \dfrac{\theta}{\theta_0} = \dfrac{25\,cm}{f}$
Total angular magnification of a microscope	$m \times M$
Angular magnification of a telescope	$M = \dfrac{\theta_e}{\theta_0} = -\dfrac{f_o}{f_e}$
Resolution	$D \sin \theta = 1.22\,\lambda$

END-OF-CHAPTER QUESTIONS

1. An object is placed 6.5 cm to the left of a 10-cm focal length lens.
   a. Draw a ray-tracing diagram that shows the image.
   b. Calculate the position and magnification of the image. Is the image real or virtual?
2. An object is placed 18 cm to the left of a 10-cm focal length lens.
   a. Draw a ray-tracing diagram that shows the image.
   b. Calculate the position and magnification of the image. Is the image real or virtual?
3. A magnifier has an angular magnification of 3.0. An object is placed 10 cm from the magnifier. Find the position and magnification of the image, and whether it is real or virtual and upright or inverted.
4. A 10× objective lens produces a real image 180 mm from the lens. The lens aperture has a diameter of 6.3 mm.
   a. Find the object's position.
   b. How close can two features be and still be resolved? Assume a wavelength of 500 nm.
5. A telescope has an aperture diameter 28 cm, objective focal length 1.0 m, and eyepiece focal length 2.5 cm.
   a. A pair of stars are 2.0 million light years (ly) away. What is the smallest distance between the two stars such that they are resolved? Assume a wavelength of 550 nm. (1 ly is the distance light travels in 1 year. Keep your answer in units of ly.)
   b. Suppose the stars are 800 ly apart. What angle is subtended by the image of the two stars?

ADDITIONAL PROBLEMS

1. Two identical lenses ($f = 30$ cm) are sandwiched together to form a compound lens. The distance between the lenses is negligible. An object is 20 cm from the compound lens. Find the position and magnification of the final image.
2. Lens 1 and lens 2 have focal lengths $f_1 = 16$ mm and $f_2 = 50$ mm (Figure 18.9). An object is 18 mm to the left of lens 1. Find the position (relative to lens 2) and magnification of the final image.

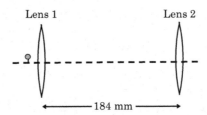

FIGURE 18.9   An object and two lenses.

3. In a confocal microscope, the object (e.g., a fluorescing cell) is at the focal point of the objective (Figure 18.10). A detector is in the focal plane of the "tube lens." Draw a ray-tracing diagram.

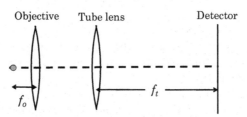

FIGURE 18.10   A confocal microscope.

4. The lenses in this chapter have been *convex*. A *concave* lens has a negative f. Rather than converge, a convex lens causes parallel rays to spread out from f (Figure 18.11). An object is placed 14 mm from concave lens with $f = -10$ mm. Using the lens equation, find the position and magnification of the image.

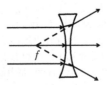

FIGURE 18.11   A concave lens.

Relativity

RELATIVE FRAMES

Suppose someone is walking along the aisle of an airplane toward the cockpit. The airplane travels at 250 m/s relative to the ground. From the airplane's *frame of reference*, the person's speed is 1.0 m/s. To an observer on the ground, the person's speed is $250 + 1 = 251$ m/s.

The velocity of an object in its "moving frame" is u'. To an observer on the ground, the object's velocity is

$$u = u' + v$$

where v is the velocity of the moving frame. In the airplane example, $u' = 1.0$ m/s and $v = 300$ m/s.

Simple addition, or "Galilean relativity," is fine for most common velocities. When speeds approach the speed of light, we must use Einstein's theory of special relativity, discussed in the remainder of this chapter.

1. An aircraft carrier travels north, 14 m/s relative to the shore. A sailor on the deck jogs south at 3.0 m/s relative to the carrier. To an observer on the shore, what is the sailor's velocity?

    $u' = -3.0$ m/s

    $v = 14$ m/s

    $u = -3 + 14 = $ **11 m/s, north**

2. A bus travels 28 m/s relative to the ground. The distance from the back of the bus to the

front is 10 m. A person walks from the back to the front in 5.0 s. To an observer on the ground, what is the person's speed?

$u' = 10/5 = 2$ m/s

$v = 28$ m/s

$u = 2 + 28 = \textbf{30 m/s}$

SPECIAL RELATIVITY

According to special relativity, the speed of light is constant in all frames. In air or vacuum, $c = 3.0 \times 10^8$ m/s.

Suppose a person is in a spaceship traveling at $0.5c$ relative to Earth. The person turns on a light. The light wave travels at a speed, c. According to Galilean relativity, an observer on the Earth would see $u = 0.5c + c = 1.5c$. But this would violate special relativity! In fact, the Earth observer would measure the speed of light as c. As discussed in the next two sections, space and time must "warp" to ensure the speed of light is constant.

1. A spaceship travels at 99% the speed of light along the x direction, relative to an observer in a space station. The spaceship shoots a laser pulse forward. According to the observer, what is the difference between the speeds of the laser and the spaceship?

   The speed of light is c.
   The speed of the spaceship is $0.99c$.
   The difference between these speeds is $c - 0.99c = \textbf{0.01c} = \textbf{3.0} \times \textbf{10}^6$ **m/s**

2. In the previous problem, what is the speed of the laser pulse according to an observer on the spaceship?

   The speed of light is $c = \textbf{3.0} \times \textbf{10}^8$ **m/s**

QUIZ 19.1

1. A satellite travels at 7286 m/s relative to an observer on Earth. A bolt pops off and flies backward (opposite to the satellite's motion). In the satellite's frame, the bolt travels at 5.0 m/s. According to the Earth observer, what is the speed of the bolt?
2. An alien ship travels at 1.6×10^8 m/s relative to an observer on Earth. The ship sends radio waves in the backward direction. According to the Earth observer, what is the speed of the radio waves?
3. A child in a car throws a paper airplane from the back to the front. In the car's frame, the paper airplane's speed is 1.5 m/s. The car travels

20.0 m/s relative to the ground. According to the ground frame, what is the speed of the paper airplane?

4. An alien in a spaceship shoots a laser pulse from the back to the front. The spaceship travels 2.8×10^8 m/s relative to a space station. According to an observer on the space station, what is the speed of the laser pulse?

LENGTH CONTRACTION

An object that is moving (relative to the ground frame) is contracted, or shortened, along the direction of motion. This does not happen because of some force that compresses the object. Space itself is contracted!

Let L' = the length of the object in its moving frame, which travels at a speed, v. L' is its "natural," normal length. The length measured by the ground frame is

$$L = L'/\gamma$$

where

$$\gamma = \frac{1}{\sqrt{1 - v^2/c^2}}$$

For any speed, v, below the speed of light, $\gamma \geq 1$. A speed greater than c would yield a negative number in the radical. Therefore, nothing can travel faster than the speed of light.

1. A meter stick ruler flies through space at 98% the speed of light, relative to Earth (Figure 19.1). According to Earth observers, what is the length of the meter stick?

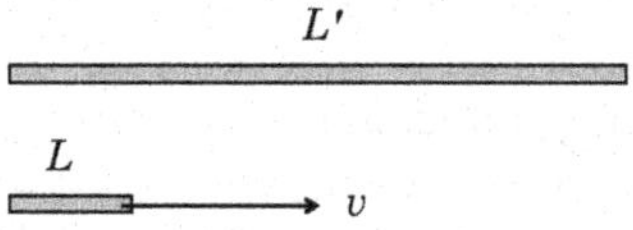

$$L'$$

$$L$$

$$\rightarrow v$$

FIGURE 19.1   A moving meter stick is length-contracted (shortened).

$$\gamma = \frac{1}{\sqrt{1 - 0.98^2}} = \frac{1}{0.20} = 5.0$$

$$L' = 1.0 \text{ m}$$

$$L = 1/5 = \mathbf{0.20 \ m}$$

2. The galaxy has a diameter of 6.0×10^{20} m. A spaceship travels at $0.9999997c$ (Figure 19.2). According to observers on the ship, how long does it take to go from one end of the galaxy to the other?

FIGURE 19.2   In the ship's frame, the length L of the galaxy is contracted.

According to the ship, the galaxy is moving toward it at $v = 0.9999997c$.

$$\gamma = \frac{1}{\sqrt{1-0.9999997^2}} = 1291$$

$L' = 6.0 \times 10^{20}$ m

$L = 6.0 \times 10^{20}/1291 = 4.6 \times 10^{17}$ m

$t = L/v = 4.6 \times 10^{17}/3.0 \times 10^8 = \mathbf{1.5 \times 10^9}$ **s** (about 50 years)

TIME DILATION

Time is measured differently by different frames. Let $\Delta t' =$ time measured by a clock in its moving frame. $\Delta t'$ is the "proper," normal time. The time measured by the ground frame is

$$\Delta t = \gamma \, \Delta t'$$

Notice how $\Delta t \geq \Delta t'$. The ground frame measures more time going by than the proper time. In other words, *moving clocks run slowly*. This is not because there is something wrong with the clock. Time itself has been stretched out, or dilated.

1. A clock moves at 98% the speed of light relative to the ground frame. The clock measures one hour (i.e., the minute hand goes around once). According to observers in the ground frame, how much time elapsed?

$$\gamma = \frac{1}{\sqrt{1-0.98^2}} = \frac{1}{0.20} = 5.0$$

$$\Delta t' = 1 \text{ hr}$$

$$\Delta t = (5)(1) = \mathbf{5\ hr}$$

(Observers in the ground frame would see the minute hand going around slowly.)

2. Albert is on a spaceship that travels $0.99c$ relative to Earth. When he begins the trip, Albert is 20 years old. His twin Isaac remains on Earth. After 10 years in the spaceship frame, Albert returns to Earth. Albert is now 30 years old. How old is Isaac?

$$\gamma = \frac{1}{\sqrt{1-0.99^2}} = \frac{1}{0.14} = 7.1$$

$$\Delta t' = 10 \text{ years}$$

$$\Delta t = (7.1)(10) = 71 \text{ years}$$

Isaac's age is $20 + 71 = \mathbf{91\ years}$

QUIZ 19.2

1. A moving object is observed to have 99% of its natural length. How fast is the object moving (as a fraction of c)?
2. Alpha Centauri is 4.2×10^{16} m from the sun (in the Earth or sun frame). A spaceship is traveling at $0.97c$. In the spaceship's frame, how long does it take to go from the sun to Alpha Centauri?
3. In the previous question, according to an observer on Earth, how long did it take for the spaceship to reach Alpha Centauri?
4. An unstable particle has a lifetime of 2.0×10^{-6} s. The particle moves at $0.87c$ relative to a laboratory. In the laboratory frame, what is the particle's lifetime?

ENERGY

A force causes a mass to accelerate. However, nothing can go faster than the speed of light. How is the "ultimate speed limit" enforced? Einstein showed us the way by changing the definition of kinetic energy.

The energy of an object is

$$E = \gamma m c^2$$

When the object is at rest, $\gamma = 1$. The *rest energy* is

$$E_0 = m c^2$$

This is a new kind of energy, which comes purely from the mass of the object. The kinetic energy, K, is the object's energy minus its rest energy:

$$K = (\gamma - 1)mc^2$$

As we do work on the object, K increases, which means γ increases. Even if γ gets huge, though, v never exceeds c.

1. What is the rest energy of a 1.0-mg object?

$$E_0 = mc^2 = (10^{-6})(9 \times 10^{16}) = \mathbf{9.0 \times 10^{10}\,J}$$

2. Calculate the kinetic energy of a proton (mass 1.67×10^{-27} kg) with a speed of $0.20c$, and compare with $\frac{1}{2}mv^2$.

$$\gamma = \frac{1}{\sqrt{1-0.2^2}} = \frac{1}{0.9798} = 1.0206$$

$$K = (0.0206)(1.67 \times 10^{-27})(9 \times 10^{16}) = \mathbf{3.10 \times 10^{-12}\,J}$$

$$\tfrac{1}{2}mv^2 = \tfrac{1}{2}(1.67 \times 10^{-27})(3.6 \times 10^{15}) = \mathbf{3.01 \times 10^{-12}\,J}$$

MOMENTUM

Like energy, momentum also has a new formula:

$$p = \gamma mv$$

As we exert force on the object, p increases. Even if γ gets huge, though, v never exceeds c.

1. Calculate the momentum of an electron (mass 9.11×10^{-31} kg) with a speed of $0.80c$.

$$\gamma = \frac{1}{\sqrt{1-0.8^2}} = \frac{1}{0.6} = 1.67$$

$$p = (1.67)(9.11 \times 10^{-31})(2.4 \times 10^8) = \mathbf{3.7 \times 10^{-22}\,kg\ m/s}$$

2. A particle with energy 6.0×10^{-13} J travels at very close to the speed of light. What is its momentum?

$$p = (\gamma m)c$$

$$= (E/c^2)c$$

$$p = E/c$$

$$= (6 \times 10^{-13})/(3 \times 10^8) = \mathbf{2.0 \times 10^{-21}\ kg\ m/s}$$

QUIZ 19.3

1. What is the rest energy of a proton?
2. Calculate the kinetic energy of an electron traveling at $0.96c$ and compare with $\frac{1}{2}mv^2$.
3. Calculate the momentum of a proton traveling at $0.40c$.
4. A particle with kinetic energy 2.9×10^{-10} J travels at close to the speed of light. What is its momentum?

CHAPTER SUMMARY

Speed of light	$c = 3.0 \times 10^8$ m/s
	Constant in all reference frames
Length contraction	$L = L'/\gamma$
	$\gamma = \dfrac{1}{\sqrt{1 - v^2/c^2}}$
Time dilation	$\Delta t = \gamma\, \Delta t'$
Energy	$E = \gamma mc^2$
Rest energy	$E_0 = mc^2$
Kinetic energy	$K = (\gamma - 1)mc^2$
Momentum	$p = \gamma mv$

END-OF-CHAPTER QUESTIONS

1. A boat travels at 9.0 m/s relative to the shore. A person throws a baseball from the back of the boat to a friend who is toward the front. The friend catches the ball. In the boat's frame, the baseball's speed was 20 m/s and the distance between the friends was 5.0 m.

   a. In the boat's frame, how long was the ball in the air?
   b. In the shore's frame, how long was the ball in the air?
   c. In the shore's frame, what was the speed of the ball?

2. A spaceship travels at $0.98c$ relative to a space station. A laser shoots a pulse forward and the pulse is absorbed by a detector. In the spaceship's frame, the distance between the laser and detector is 100 m.
   a. In the spaceship's frame, how long did it take for the pulse to reach the detector?
   b. In the space station's frame, what was the distance between the laser and detector?
   c. In the space station's frame, what was the speed of the laser pulse?

3. A space probe with a clock takes 50.0 years to go from Earth to Alpha Centauri, a distance of 4.2×10^{16} m (Earth's frame). According to the probe's clock, how many years did it take? (1 year $= 3.16 \times 10^7$ s)

4. An electron travels at 2.0×10^8 m/s. Calculate its
   a. Rest energy
   b. Energy
   c. Kinetic energy
   d. Momentum

5. A powerful force does 2.7×10^{14} J of work on a 1.0-g object that starts from rest.
   a. What would the object's speed, v, be, if $K = \frac{1}{2}\, mv^2$?
   b. Calculate γ.
   c. From part (b), calculate the object's speed.

ADDITIONAL PROBLEMS

1. A spaceship travels from Earth to the end of the galaxy and back, for a total distance of 1.0×10^{21} m. The spaceship's speed is very close to the speed of light. (The spaceship would need time to accelerate from rest and also turn around, but we can ignore those details.)
   a. How many years elapsed on Earth?
   b. The ship's clock only recorded 5 years. Calculate γ.

2. A spaceship chases an enemy spaceship. The two spaceships travel at $0.98c$ relative to Planet X. The enemy spaceship fires a laser pulse toward the rear. In the spaceships' frame, the pulse travels 300 m before hitting the chasing spaceship. In the planet's frame,
   a. What was the distance between the spaceships?
   b. How long did it take for the pulse to hit the spaceship? (Note: The two events, firing the pulse and the pulse hitting the spaceship, occur at different locations. Because of this, we cannot use $\Delta t = \gamma \Delta t'$).

3. A positron is a positively charged particle with a mass of 9.11×10^{-31} kg. Consider a positron that travels with a velocity $v_x = 2.7 \times 10^8$ m/s and collides with a proton. After the collision, the speed of the positron is barely reduced, but its direction reverses (i.e., $v_x = -2.7 \times 10^8$ m/s). What is the velocity of the proton after the collision? (Hint: The proton speed is low, such that $p = mv$ is a good approximation.)

4. An electron starts from rest and is accelerated by a 1.7×10^5 V/m electric field over a distance of 6.0 m. What is its final speed?

Atoms and Nuclei

THE ELECTRON VOLT

In atomic and nuclear physics, energies are often expressed as eV, or "electron volt." Recall that the potential energy (PE) is

$$PE = qV$$

where q is the charge, and V is the potential. As such, 1 eV is defined as the potential energy of a $q = 1.6 \times 10^{-19}$ C charge at a potential of 1 V:

$$1 \text{ eV} = (1.6 \times 10^{-19} \text{ C})(1 \text{ V}) = 1.6 \times 10^{-19} \text{ J}$$

Common prefixes are

$$1 \text{ keV} = 10^3 \text{ eV} = 1.6 \times 10^{-16} \text{ J}$$

$$1 \text{ MeV} = 10^6 \text{ eV} = 1.6 \times 10^{-13} \text{ J}$$

1. What is the speed of an electron with a kinetic energy of 60 eV?

   This energy is nonrelativistic, so $K = \frac{1}{2} mv^2$

   $\frac{1}{2} mv^2 = 60 \ (1.6 \times 10^{-19} \text{ J})$

   $(0.5)(9.1 \times 10^{-31})v^2 = 9.6 \times 10^{-18}$

   $v^2 = 2.1 \times 10^{13}$

   $v = 4.6 \times 10^6 \text{ m/s}$

2. An electron starting from rest is accelerated from a negative plate to a positive plate. The potential difference between the plates is 1.2 kV. What is the kinetic energy of the electron before it hits the positive plate?

$$\Delta PE = (-1.6 \times 10^{-19})(1.2 \times 10^3) = -1.2 \times 10^3 \, eV = -1.2 \, keV$$

$$\Delta KE = -\Delta PE = \textbf{1.2 keV}$$

THE BOHR MODEL

An atom consists of electrons orbiting around a positive nucleus. The simplest atom is hydrogen. Its nucleus is a proton. One electron orbits around the nucleus, similar to a planet orbiting a star (Figure 20.1).

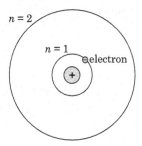

FIGURE 20.1 Bohr model of the hydrogen atom.

Unlike a planetary orbit, the electron's total energy E (kinetic plus potential) is *quantized*. It can only have specific values, given by

$$E = -\frac{13.6 \, eV}{n^2}$$

where n is a positive integer (1, 2, 3, ...). An n value corresponds to an orbit of radius

$$r = (0.53 \, \text{Å})n^2$$

where 1 Angstrom (Å) $= 10^{-10}$ m. The *ground state* ($n = 1$) has an orbit of radius 0.53 Å and energy −13.6 eV. Higher n values correspond to larger orbits and less-negative energies.

1. What is the energy and orbital radius of an $n = 2$ electron in a hydrogen atom?

$$E = -13.6/4 = \textbf{−3.4 eV}$$

$$r = (0.53)(4) = \textbf{2.1 Å}$$

2. How much energy does it take to remove a ground-state electron from a hydrogen atom?

If $E \geq 0$, then the electron is not bound to the hydrogen atom.

$E(\text{before}) = -13.6$ eV

$E(\text{after}) = 0$

$E(\text{after}) - E(\text{before}) = \mathbf{13.6\ eV}$

QUIZ 20.1

1. What is the speed of a proton with a kinetic energy of 60 eV?
2. An electron travels from a region at 5.0 V to a region at 2.0 V. What is the change in kinetic energy?
3. What is the energy and orbital radius of an $n = 4$ electron in a hydrogen atom?
4. How much energy does it take to remove an $n = 2$ electron from a hydrogen atom?

PHOTONS

In Chapter 17, we learned that light is an electromagnetic wave. According to quantum mechanics, light energy is quantized (i.e., the energy can only assume discrete values). A *photon* is a single "quantum" of energy and can be thought of as a particle. Its energy is

$$E = hf$$

where $h = 6.6 \times 10^{-34}$ J·s, and f is the frequency. The wave speed equation is the same as before:

$$c = \lambda f$$

where c is the speed of light, and λ is wavelength.

When a photon is *absorbed* by a hydrogen atom, the photon is destroyed and its energy goes to increase the electron's energy. If an electron loses energy (e.g., by going from $n = 3$ to 2), it can *emit* a photon with that energy. In this way, energy is conserved.

1. A photon is absorbed by a hydrogen atom, causing the electron to go from $n = 1$ to $n = 2$. What is the photon's energy, frequency, and wavelength?

$E(\text{before}) = -13.6/1 = -13.6 \text{ eV}$

$E(\text{after}) = -13.6/4 = -3.4 \text{ eV}$

The photon energy is $E(\text{after}) - E(\text{before}) = \textbf{10.2 eV}$

$$E = (10.2)(1.6 \times 10^{-19}) = 1.63 \times 10^{-18} \text{ J}$$

$E = hf$

$$1.63 \times 10^{-18} = (6.6 \times 10^{-34})f$$

$$\textbf{2.5} \times \textbf{10}^{\textbf{15}} \textbf{ Hz} = f$$

$c = \lambda f$

$$3 \times 10^8 = \lambda(2.5 \times 10^{15})$$

$$1.2 \times 10^{-7} \text{ m} = \lambda, \text{ or } \textbf{120 nm}$$

2. An electron transitions from $n = 4$ to $n = 3$ and emits a photon. What is the photon's energy, frequency, and wavelength?

$E(\text{before}) = -13.6/16 = -0.85 \text{ eV}$

$E(\text{after}) = -13.6/9 = -1.51 \text{ eV}$

$E(\text{after}) - E(\text{before}) = -0.66 \text{ eV}$

To conserve energy, a photon of energy **0.66 eV** is emitted.

$$E = (0.66)(1.6 \times 10^{-19}) = 1.06 \times 10^{-19} \text{ J}$$

$E = hf$

$$1.06 \times 10^{-19} = (6.6 \times 10^{-34})f$$

$$\textbf{1.6} \times \textbf{10}^{\textbf{14}} \textbf{ Hz} = f$$

$c = \lambda f$

$$3 \times 10^8 = \lambda(1.6 \times 10^{14})$$

$$1.87 \times 10^{-6} \text{ m} = \lambda, \text{ or } \textbf{1870 nm}$$

NUCLEI

The atomic nucleus consists of protons and neutrons (Figure 20.2). The nucleus is about 10,000 times smaller than the electrons' orbital diameter.

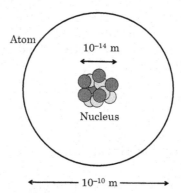

FIGURE 20.2   Model of the atom (not to scale).

The number of protons is Z, also called the "atomic number." The nucleus has a charge of $+Zq$, where $q = 1.6 \times 10^{-19}$ C. If the atom is neutral, then there are also Z electrons—their negative charge balances the protons' positive charge. Each Z number corresponds to a specific element: $Z = 1$ is hydrogen, $Z = 2$ is helium, etc.

The number of neutrons is N. A neutron is a neutral particle (no charge) with a mass slightly greater than a proton. Neutrons do not affect the chemical properties of the element significantly. *Isotopes* have the same Z but different N.

The number of protons plus neutrons, called the "mass number," is $A = Z + N$. An isotope is designated by the element symbol with A as a leading superscript. For example, carbon-12 ($Z = 6$, $N = 6$) and carbon-13 ($Z = 6$, $N = 7$) are written as ^{12}C and ^{13}C.

1. Lithium ($Z = 3$) has two common isotopes, ^{6}Li and ^{7}Li. For each isotope, list the number of protons, neutrons, and electrons, assuming the atom is neutral.

   ^{6}Li: **3 protons, 3 neutrons, 3 electrons**
   ^{7}Li: **3 protons, 4 neutrons, 3 electrons**

2. List the number of protons, neutrons, and electrons in ^{14}C ($Z = 6$) and ^{14}N ($Z = 7$).

   ^{14}C: **6 protons, 8 neutrons, 6 electrons**
   ^{14}N: **7 protons, 7 neutrons, 7 electrons**

QUIZ 20.2

1. An electron in a hydrogen atom has $n = 2$. What frequency of light will promote the electron to $n = 3$?
2. A hydrogen electron with $n = 2$ decays to the ground state by emitting a photon. What is the wavelength of the photon?
3. How many neutrons are in ^{208}Pb $(Z = 82)$?
4. List the number of protons, neutrons, and electrons in neutral ^{16}N $(Z = 7)$ and ^{16}O $(Z = 8)$.

ATOMIC MASS

Atomic mass is measured in units of u, where

$$u = 1.660539 \times 10^{-27}\,\text{kg}$$

Protons and neutrons in the nucleus are bound together by the *strong force*. This force lowers the total energy of the atom. Because $E_0 = mc^2$, this is equivalent to saying the atomic mass is lowered.

A nuclear reaction can occur if it results in a decrease in mass. This is like an exothermic reaction in chemistry. The nuclear reaction energy is

$$E = (-\Delta m)c^2$$

where $-\Delta m = m(\text{initial}) - m(\text{final})$. $E > 0$ means that the reaction is exothermic and will occur eventually. If Δm is measured in atomic mass units (u),

$$E = -\Delta m \times 931.5 \text{ MeV}$$

1. Hydrogen and deuterium undergo nuclear *fusion* to form helium-3:

$$^1\text{H} + {}^2\text{H} \rightarrow {}^3\text{He}$$

Find the nuclear reaction energy. The atomic masses are:

^{1}H 1.007825

^{2}H 2.014102

^{3}He 3.016030

Note: Atomic masses are listed for neutral atoms, so they include the electron masses.

$m(\text{initial}) = 1.007825 + 2.014102 = 3.021927$

$m(\text{final}) = 3.016030$

$-\Delta m = 3.021927 - 3.016030 = 5.897 \times 10^{-3}$

$E = 5.897 \times 10^{-3} \times 931.5 = \mathbf{5.5\ MeV}$

2. In α *decay*, ^{4}He is ejected from the nucleus. Because ^{4}He has 2 protons and 2 neutrons, this process reduces Z by 2 and A by 4. Find the missing isotope and the reaction energy:

$$^{210}\text{Po} \rightarrow \underline{\quad} + {}^4\text{He}$$

Atomic masses are listed here for Po ($Z = 84$), Pb ($Z = 82$), and He ($Z = 2$):

^{210}Po 209.982848

^{206}Pb 205.974440

^{207}Pb 206.975872

^{208}Pb 207.976627

^{4}He 4.002603

After decay, $Z = 84 - 2 = 82$, which is Pb

$A = 210 - 4 = 206$, so the isotope is $\mathbf{^{206}Pb}$

$m(\text{initial}) = 209.982848$

$m(\text{final}) = 205.974440 + 4.002603 = 209.977043$

$-\Delta m = 209.982848 - 209.977043 = 5.805 \times 10^{-3}$

$E = 5.805 \times 10^{-3} \times 931.5 = \mathbf{5.4\ MeV}$

RADIOACTIVITY

If there is a nuclear reaction with $E > 0$, then it will occur, sooner or later. The nucleus will decay into a "daughter" nucleus. An isotope that can undergo such a decay is unstable, or *radioactive*.

Radioactivity is a random process. The *half-life* ($t_{1/2}$) is the time it takes, on average, for 50% of the radioactive nuclei to decay. The number of nuclei that have not yet decayed is

$$N = N_0 e^{-t/\tau}$$

where N_0 is the number of nuclei at $t = 0$ and

$$\tau = t_{1/2}/0.693$$

Activity (in units called Becquerel, or Bq) is the number of decays per second:

$$R = N/\tau$$

1. Carbon-14 has a half-life of 5730 years. In a living organism the abundance of ^{14}C is 1.3×10^{-12} (i.e., a small fraction of carbon atoms is carbon-14). What is the ^{14}C abundance 10,000 years after the organism dies?

   We can let N denote the abundance.

   $\tau = 5730/0.693 = 8270$ years

   $N = (1.3 \times 10^{-12})\, e^{-10,000/8270} = \mathbf{3.9 \times 10^{-13}}$

2. A rock contains 1.0×10^{22} uranium-238 atoms. ^{238}U has a half-life of 4.5×10^9 years (1 year $= 3.16 \times 10^7$ s). What is the activity?

   $\tau = 4.5 \times 10^9/0.693 = 6.5 \times 10^9$ years

   $(6.5 \times 10^9 \text{ years})(3.16 \times 10^7 \text{ s/year}) = 2.05 \times 10^{17}$ s

   $R = (1.0 \times 10^{22})/(2.05 \times 10^{17}) = \mathbf{4.9 \times 10^4\, Bq}$

QUIZ 20.3

1. Deuterium and tritium undergo nuclear fusion to form helium-4 and a neutron:

   $$^2\text{H} + {}^3\text{H} \rightarrow {}^4\text{He} + \text{n}$$

   Find the nuclear reaction energy. The atomic masses are:

   ^{2}H 2.014102

   ^{3}H 3.016050

^{4}He 4.002603

n   1.008665

2. In *β decay*, a neutron decays into a proton and electron. Because a neutron is converted to a proton, Z increases by 1 but A is unchanged. Find the missing isotope and the reaction energy:

$$^{60}\text{Co} \rightarrow \underline{\ \ } + e$$

Atomic masses are listed here for Co ($Z = 27$) and Ni ($Z = 28$):

^{60}Co 59.933819

^{58}Ni 57.935346

^{60}Ni 59.930788

Note that atomic masses are listed for the *neutral* atoms. Because neutral Ni contains one more electron than neutral Co, Ni already contains the mass of the electron in the reaction equation. We do not need to add it again.

3. A specimen contains 5.0×10^{-6} mole of ^{31}P ($t_{1/2} = 2.6$ hours). How many moles of ^{31}P will there be after 24 hours?

4. A banana contains 7.2×10^{17} potassium-40 atoms. The half-life of ^{40}K is 1.25×10^9 years. What is the activity?

CHAPTER SUMMARY

Electron volt	$1\ \text{eV} = 1.6 \times 10^{-19}\ \text{J}$
Hydrogen atom	$E = -\dfrac{13.6\ \text{eV}}{n^2}$
	$r = (0.53\ \text{Å})n^2$
Photon	$E = hf$
	$h = 6.6 \times 10^{-34}\ \text{J·s}$
	$c = \lambda f$
Atomic mass unit	$u = 1.660539 \times 10^{-27}\ \text{kg}$
Nuclear reaction	$E = -\Delta m \times 931.5\ \text{MeV}$
energy	$-\Delta m = m(\text{initial}) - m(\text{final})$
Nuclear decay	$N = N_0 e^{-t/\tau}$
	$\tau = t_{1/2}/0.693$
Activity	$R = N/\tau$

END-OF-CHAPTER QUESTIONS

1. What is the energy and orbital radius of an electron in a hydrogen atom with $n = 23$?
2. An electron in a hydrogen atom is excited from the ground state to the $n = 4$ state.
   a. How much energy (eV) was required to excite the electron?
   b. The electron goes from $n = 4$ to $n = 3$. A photon is emitted. What is its frequency?
   c. What is the wavelength of the emitted photon?
3. ^{239}Pu ($Z = 94$) is an unstable isotope that decays into ^{235}U ($Z = 92$).
   a. How many protons, neutrons, and electrons are in a neutral ^{239}Pu atom?
   b. What is the nuclear reaction energy? Atomic masses are listed here:

   ^{235}U   235.043924

   ^{239}Pu 239.052157

   ^{4}He   4.002603

4. Phosphorus-32 undergoes β decay. What is the nuclear reaction energy? Atomic masses are listed here for P ($Z = 15$) and S ($Z = 16$):

   ^{32}P 31.973907

   ^{32}S 31.972070

   ^{35}S 34.969031

5. A sample contains 4.6×10^{16} germanium-71 atoms, which have a half-life of 11.4 days.
   a. What is the activity?
   b. How many ^{71}Ge atoms will there be after 7.0 days?

ADDITIONAL PROBLEMS

1. A hydrogen atom is in the $n = 2$ state. A photon of wavelength 266 nm is absorbed.
   a. What was the photon energy (eV)?
   b. Using energy conservation, find the energy of the electron after absorbing the photon.

2. The energy of a single electron orbiting a nucleus with Z protons is

$$E = -Z^2 \frac{13.6 \text{ eV}}{n^2}$$

How much energy is required to liberate a ground-state electron from He^+?
3. *Electron capture* is a process where an atomic electron is captured by a proton, converting it to a neutron. Find the nuclear reaction energy for the electron capture decay of ^{65}Zn. Some masses are listed here for Cu ($Z = 29$), Zn ($Z = 30$), and Ga ($Z = 31$):

^{65}Cu 64.927793

^{65}Zn 64.929241

^{65}Ga 64.932735

4. An isolated neutron (one that is not in an atom) decays into a proton and electron with a half-life of 10 minutes:

$$n \rightarrow p + e$$

a. What would be the activity of 73,800 neutrons?
b. What is the nuclear reaction energy for neutron decay? Masses:

n 1.008665

p 1.007276

e 0.000549

Answers

CHAPTER 1

QUIZ 1.1

1. Displacement = 45 m, velocity = 5.0 m/s
2. Displacement = −40 cm, velocity = −8.0 cm/s
3. 50 m/s^2
4. −4.0 m/s^2

QUIZ 1.2

1. 0.0 m/s
2. 30 m/s
3. 19 m
4. −5.0 m

QUIZ 1.3

1. 2.0 s
2. 10 s
3. 3.0 s
4. 20 s

END-OF-CHAPTER QUESTIONS

1. (a)−4.0 m/s^2, (b) 8.0 m
2. 500 m
3. 10 s
4. 5.0 s
5. 7.1 s

ADDITIONAL PROBLEMS

1. (a) 1.0 s, (b) 4.0 m/s
2. (a) 5.0 s, (b) 300 m
3. (a) 1.3 s, (b) –13 m/s
4. 0.72 s

CHAPTER 2

QUIZ 2.1

1. $v_x = 2.0$ m/s, $v_y = -4.0$ m/s
2. $v = 13$ m/s, $\theta = 23°$
3. $v_y = 75$ m/s
4. 150°

QUIZ 2.2

1. $C_x = -2.5$, $C_y = 2.0$
2. 53 m/s, 20° south of east
3. (–3.7 m, 6.0 m)
4. (–390 m, 40 m)

QUIZ 2.3

1. 4.0 s
2. $v_x = 34$ m/s, $v_y = -40$ m/s
3. 2.0 s
4. 8.3 m

END-OF-CHAPTER QUESTIONS

1. $v_x = 24$ m/s, $v_y = 97$ m/s
2. $C_x = 4.0$, $C_y = 4.3$
3. (a) 3.0 s, (b) $v_x = 7.0$ m/s, $v_y = -30$ m/s
4. (a) 4.0 s, (b) 80 m
5. (a) 0.51 s, (b) 2.8 m

ADDITIONAL PROBLEMS

1. 2.0 s
2. 1.0 s

3. (a) 5.1 s, (b) 44 m
4. 3.8 m

CHAPTER 3

QUIZ 3.1

1. $F_x = -2.5$ N, $F_y = 4.0$ N
2. 0.80 m/s^2
3. $F_y = -78$ N, weight = 78 N
4. $a_y = -9.8$ m/s^2 (both)

QUIZ 3.2

1. $F_x = 5.0$ N
2. $F_y = -10$ N
3. 2900 N
4. 0.050 N

QUIZ 3.3

1. 9.8 N, up
2. 380 N, up
3. 290 N
4. 2.6 m/s^2

END-OF-CHAPTER QUESTIONS

1. $a_x = 0.50$ m/s^2, $a_y = -0.25$ m/s^2
2. $F_x = -10,300$ N
3. 5.2 N, up
4. 15 N
5. (a) 1500 N, (b) 3.7 m/s^2

ADDITIONAL PROBLEMS

1. (a) 980 N, up, (b) 1000 N, +x direction
2. 140 N
3. 6.1 s
4. (a) 1.0 N, (b) 2.9 m/s^2

CHAPTER 4

QUIZ 4.1

1. 39 N
2. 2.0 m/s^2
3. 22 N, up
4. 0.88 N (+x direction)

QUIZ 4.2

1. $a_x = 6.9$ m/s^2
2. 28 N
3. 22°
4. $a_x = 7.5$ m/s^2

QUIZ 4.3

1. $a_y = -7.8$ m/s^2
2. 29 N
3. $a_y = -5.9$ m/s^2
4. 160 N

END-OF-CHAPTER QUESTIONS

1. 20 N
2. (a) $a_x = 0.19$ m/s^2, (b) 960 N
3. (a) $a_x = 4.9$ m/s^2, (b) $a_x = 2.8$ m/s^2
4. 98 N
5. (a) 49 N, (b) $a_y = 3.3$ m/s^2

ADDITIONAL PROBLEMS

1. (a) 470 N, up, (b) 12 m
2. (a) $a_x = -5.0$ m/s^2, (b) $a_x = -10$ m/s^2, $a_y = 7.5$ m/s^2
3. 1170 N
4. $a_x = -1.7$ m/s^2

CHAPTER 5

QUIZ 5.1

1. 1.0 m/s^2, toward center
2. 5900 m/s^2, toward center
3. 130 N
4. 9.8 N, up

QUIZ 5.2

1. 50 N, toward center
2. 9.9 m/s
3. 67 N
4. 250 N

QUIZ 5.3

1. 5.1 × 10^{-7} N
2. 11 m/s^2
3. 3.0 × 10^4 m/s
4. 4.0 × 10^5 s

END-OF-CHAPTER QUESTIONS

1. 0.016 N
2. 34 m/s
3. 89 N
4. 2900 N
5. speed = 3900 m/s, period = 4.3 × 10^4 s

ADDITIONAL PROBLEMS

1. 70 m/s
2. (a) 8.1 N, (b) 0.32 N
3. 71 N
4. 420 m

CHAPTER 6

QUIZ 6.1

1. 14 J
2. 250 J
3. −250 J
4. 4.9×10^5 J

QUIZ 6.2

1. 8.0 m/s
2. −40 J
3. 20 m/s
4. 5.1 m

QUIZ 6.3

1. 1.0×10^{-3} J
2. 2.0×10^{-2} m/s
3. 1240 J
4. 5.7 m/s

END-OF-CHAPTER QUESTIONS

1. 1.2×10^5 m/s
2. −15 J
3. 6.3 m/s
4. (a) 5.0 m/s, (b) 15 m/s
5. 5.9×10^5 J

ADDITIONAL PROBLEMS

1. −30 J
2. 7.5 m/s
3. 1400 W
4. 150 W

CHAPTER 7

QUIZ 7.1

1. 7.3×10^{-5} rad/s
2. 470 m/s
3. 2.6×10^{29} J
4. 9.1×10^{-3} J

QUIZ 7.2

1. 0.15 kg m^2
2. 12 kg m^2
3. 4.0 N m
4. 1.2 N m

QUIZ 7.3

1. –1.3 N m
2. 61 N m
3. 0.021 N m
4. –5.0 rad/s^2

END-OF-CHAPTER QUESTIONS

1. (a) 1.6 rad/s, (b) 3.1 m/s, tangential
2. 2.6×10^5 J
3. 260 J
4. 2300 N m
5. 0.89 rad/s^2

ADDITIONAL PROBLEMS

1. (a) –88 N m, (b) 12 J
2. 17 N m
3. (a) 16 J, (b) 2.0 rad/s
4. (a) 600 kg m^2, (b) 740 J

CHAPTER 8

QUIZ 8.1

1. 1.0×10^5 kg m/s
2. 20,000 N
3. 0.50 m/s
4. 24 J

QUIZ 8.2

1. 1.0 m/s, $-x$ direction
2. 0.0 J
3. 7.1×10^{33} kg m²/s
4. -7.5×10^6 kg m²/s

QUIZ 8.3

1. 24 rad/s
2. 3.8 rad/s
3. 0.75 rad/s
4. 0.067 rad/s

END-OF-CHAPTER QUESTIONS

1. 300 N
2. (a) 1.0 m/s, (b) 12 J
3. 9.9 m/s
4. 2.2 rad/s
5. 4.0 rad/s

ADDITIONAL PROBLEMS

1. 0.033 m
2. (a) 3.0 m/s, (b) 54 J
3. 52 km/s
4. 22 rpm

CHAPTER 9

QUIZ 9.1

1. 5.1 N
2. 200 N/m
3. amplitude = 7.0 cm, frequency = 5.0 Hz, period = 0.20 s
4. 0.50 s

QUIZ 9.2

1. 25 cm/s
2. 1.3 m/s^2
3. $A = 2.0$ m, $\lambda = 20$ m
4. $y = 2.0 \cos(0.1\pi x - 0.4\pi t)$

QUIZ 9.3

1. 1200 m/s
2. 250, 500, 750 Hz
3. 1.6 m
4. path-length difference = 2.0 m, quiet

END-OF-CHAPTER QUESTIONS

1. 0.70 Hz
2. (a) 70 cm/s, equilibrium ($x = 0$); (b) 0.025 J, $x = \pm7.0$ cm
3. $A = 7.0$ cm, $\lambda = 0.50$ m, $f = 6.0$ Hz, $v = 3.0$ m/s, direction $= -x$
4. 2.0, 4.0, 6.0 Hz
5. loud

ADDITIONAL PROBLEMS

1. 1.6 s
2. $f = 1.6$ Hz, $T = 0.63$ s
3. −16 cm
4. (a) 0.050 m, (b) small

CHAPTER 10

QUIZ 10.1

1. 1.6×10^6 N
2. 62 m
3. 0.071 m
4. 754 mm

QUIZ 10.2

1. 0.049 N, up
2. 0.060 m
3. 16 m/s
4. 1.0 cm/s

QUIZ 10.3

1. 114 kPa
2. 2.33×10^5 Pa
3. 110.5 kPa
4. up

END-OF-CHAPTER QUESTIONS

1. (a) 3 atm, (b) 0.080 m^3
2. (a) 1.79×10^5 Pa, (b) 1.88×10^5 Pa
3. 2.0 m/s
4. (a) 20 m/s, (b) 2.5×10^5 Pa
5. 17.0 m/s

ADDITIONAL PROBLEMS

1. 130 N
2. 96 mm Hg
3. 6.8 mm
4. (a) 1.89 m/s, (b) 0.82 cm

CHAPTER 11

QUIZ 11.1

1. 2.7 g
2. 22 L
3. 620 J
4. pressure $= 3.0$ atm, energy $= 460$ J

QUIZ 11.2

1. 130 J
2. −120 J
3. 8.0 J (into gas)
4. 2.0×10^5 J (into gas)

QUIZ 11.3

1. 0.56 (56%)
2. 0.27 (27%)
3. 1.5×10^{-5} J/K
4. −0.020 J/K

END-OF-CHAPTER QUESTIONS

1. (a) 0.25 m³, (b) 37,400 J
2. (a) 0.082 m³, (b) 12,500 J, (c) 20,800 J
3. −250 J (out of the box)
4. (a) 0.34 (34%), (b) 17 J
5. 2.9×10^{-3} J/K

ADDITIONAL PROBLEMS

1. (a) 10 L, (b) decrease
2. 280 K
3. 0.99 mole
4. 8.0 J

CHAPTER 12

QUIZ 12.1

1. 8.2×10^{-8} N
2. $F_x = -30$ N
3. 11.3 N
4. $F_x = 5.6$ N, $F_y = 9.7$ N

QUIZ 12.2

1. 2.0×10^8 N/C
2. $E_x = 1.5 \times 10^7$ N/C
3. $E_x = 9000$ N, $E_y = 2000$ N
4. $E = 9200$ N/C, $\theta = 13°$ (above x axis)

QUIZ 12.3

1. $E_x = -8.4 \times 10^5$ N/C, $E_y = -2.2 \times 10^5$ N/C
2. $F_x = -4.2 \times 10^{-3}$ N, $F_y = -1.1 \times 10^{-3}$ N
3. $a_y = -9.6 \times 10^{11}$ m/s²
4. $v_x = 3.0 \times 10^5$ m/s, $v_y = -9.6 \times 10^5$ m/s

END-OF-CHAPTER QUESTIONS

1. 1.9×10^6
2. $F_x = -32$ N
3. 0.83 m
4. 18 N/C, $258°$ clockwise from y axis
5. $a_x = 4.8 \times 10^{11}$ m/s²

ADDITIONAL PROBLEMS

1. 2.1×10^{-10} N
2. $E_x = -1.4 \times 10^6$ N/C, $E_y = 3.9 \times 10^6$ N/C
3. $E_x = -2.8 \times 10^5$ N/C, $E_y = -9.6 \times 10^4$ N/C
4. $77°$

CHAPTER 13

QUIZ 13.1

1. -7.7×10^{-18} J
2. 9.6×10^4 m/s
3. 1500 V/m
4. 4.5 V

QUIZ 13.2

1. 1.4×10^{-14} J
2. 4.2×10^6 m/s
3. 2.0 m
4. 2700 V

QUIZ 13.3

1. See Figure 1.

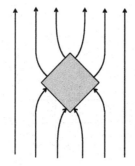

FIGURE 1   Quiz 13.3, problem 1.

2. See Figure 2.

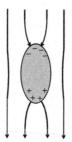

FIGURE 2   Quiz 13.3, problem 2.

3. $Q = 1.8 \times 10^{-4}$ C. See Figure 3.

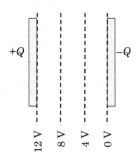

FIGURE 3   Quiz 13.3, problem 3.

4. $Q = 4.5 \times 10^{-10}$ C

END-OF-CHAPTER QUESTIONS

1. (a) 4.0×10^{21}, (b) -4.0×10^{8} V
2. 1.8×10^{6} m/s
3. 5.1 m, 36 m
4. 9.0×10^{5} V
5. (a) $Q = 9.6 \times 10^{-8}$ C, (b) See Figure 4.

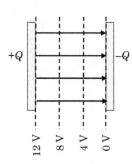

FIGURE 4   End-of-chapter questions, Chapter 13, problem 5(b).

ADDITIONAL PROBLEMS

1. 1.2×10^{4} m/s
2. 6.9 m
3. ±1.2 m
4. For a solid metal box, $\mathbf{E} = 0$ inside. This is also true for a hollow box. See Figure 5.

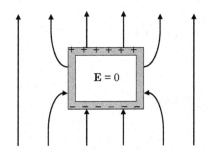

FIGURE 5   Additional problems, Chapter 13, problem 4.

CHAPTER 14

QUIZ 14.1

1. 9.4×10^{19}
2. 1.2×10^{-4} A
3. 500 W
4. 18 W

QUIZ 14.2

1. 3.6×10^{-2} A
2. 100 Ω resistors: 3.6 V each, 50 Ω resistor: 1.8 V
3. 1.9×10^{-3} A
4. 1 kΩ resistor: 1.5×10^{-3} A, 4 kΩ resistor: 3.8×10^{-4} A

QUIZ 14.3

1. 0.10 A
2. (a) 5.4 V, (b) 0 V
3. 1 MΩ resistor: 1.2×10^{-5} A, 400 kΩ resistor: 3.0×10^{-5} A, current supplied: 4.2×10^{-5} A
4. 50 Ω resistor: 2.6×10^{-2} A, 100 Ω resistor: 1.7×10^{-2} A, 200 Ω resistor: 8.6×10^{-3} A

END-OF-CHAPTER QUESTIONS

1. 1.8×10^{-4} A, 1.6×10^{-3} W
2. 2.0 W

3. 0.83 A, 7.5 W
4. (a) 0.10 A, (b) 0.30 V, (c) 0.030 W
5. 100 Ω resistor: 1.6×10^{-2} A, 400 Ω resistor: 1.1×10^{-2} A, 800 Ω resistor: 5.5×10^{-3} A

ADDITIONAL PROBLEMS

1. 7.2 W
2. closed: 5.0×10^{-4} A, open: 8.3×10^{-5} A
3. 7 Ω and 11 Ω resistors: 0.50 A, 27 Ω resistor: 0.22 A
4. 3.0×10^{-2} A

CHAPTER 15

QUIZ 15.1

1. The answer is "a".
2. See Figure 6.

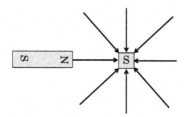

FIGURE 6   Quiz 15.1, problem 2.

3. See Figure 7.

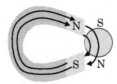

FIGURE 7   Quiz 15.1, problem 3.

4. See Figure 8.

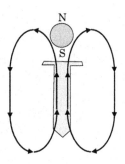

FIGURE 8   Quiz 15.1, problem 4.

QUIZ 15.2

1. 0.43 A
2. $B_x = 6.0 \times 10^{-4}$ T, $B_y = 0$
3. 3.2 A
4. 6.3×10^{-2} T

QUIZ 15.3

1. 1.1×10^{-2} N, into the page
2. 1.4×10^{13} m/s², up
3. 0.23 N, out of the page
4. 1.0×10^{-5} N, away from the other wire

END-OF-CHAPTER QUESTIONS

1. (a) 3.1×10^{-4} T, out of the page, (b) N will be out of the page, S into the page
2. (a) 5.0×10^{-2} T, (b) repelled
3. (a) 1.2×10^{-3} T, into the page, (b) 0.58 N, up
4. $B_x = 3.2 \times 10^{-5}$ T, $B_y = -1.1 \times 10^{-5}$ T
5. 4.8×10^{-18} N, down

ADDITIONAL PROBLEMS

1. 5.7×10^5 V/m
2. 1.0×10^{-5} m
3. (a) 1.0×10^{-3} N, (b) 1.7×10^{-5} N m
4. 9.4×10^{-4} N, along the axis of the motor. Torque = 0.

CHAPTER 16

QUIZ 16.1

1. 2.0×10^{-4} T m^2
2. 2.5×10^{-3} T m^2
3. Clockwise
4. Counterclockwise

QUIZ 16.2

1. 0.45 V, counterclockwise
2. 2.4 V, clockwise
3. 7.9×10^{-8} V, counterclockwise
4. 3.1 V

QUIZ 16.3

1. 10 V
2. 1500 V
3. 45
4. 0.090 A

END-OF-CHAPTER QUESTIONS

1. Clockwise
2. (a) 0.50 V, (b) Counterclockwise
3. (a) 0.0105 A, counterclockwise, (b) 3.15×10^{-4} N, toward the right
4. 1.9×10^{-3} A
5. Voltage = 600 V, current = 1.0 A

ADDITIONAL PROBLEMS

1. 0.043 V
2. 2.8×10^{-6} A, counterclockwise
3. (a) Clockwise, (b) up (the magnetic poles of the ring and solenoid repel)
4. (a) 0.18 A, clockwise, (b) 5.4×10^{-3} N, toward the right

CHAPTER 17

QUIZ 17.1

1. 3.0×10^5 m
2. 9.2×10^{14} Hz
3. 0.018 W/cm^2
4. Electric field: 0.31, intensity: 0.095

QUIZ 17.2

1. 5.0×10^{-4} m
2. 1.1°
3. 8.0×10^{-3} m
4. 6.0×10^{-3} m

QUIZ 17.3

1. Blue: 41.9°, red: 41.5°
2. 3.1 m
3. 125 nm
4. 100 nm

END-OF-CHAPTER QUESTIONS

1. (a) 1.0×10^{15} Hz, (b) 2.3×10^{-4} W/cm^2
2. (a) 0.38°, (b) 9.9×10^{-3} m, (c) 0.13 m
3. 1.1×10^{-2} m
4. (a) 2.1×10^8 m/s, (b) 10° below horizontal
5. 760 nm, 591 nm, 484 nm

ADDITIONAL PROBLEMS

1. (a) 8.5×10^{-3} m, (b) 0.026 W/cm^2
2. 1.6×10^{-5} m
3. 42°
4. 1.3

CHAPTER 18

QUIZ 18.1

1. See Figure 9.

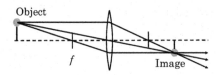

FIGURE 9   Quiz 18.1, problem 1.

2. $s' = 17$ cm, $m = -0.67$, real image
3. $s' = 930$ cm, $m = -30$
4. $s' = -20$ cm (same side as object), size = 16 mm, upright

QUIZ 18.2

1. $s' = -2.5$ cm, $m = 1.25$
2. 35°
3. $m = -50$
4. 125

QUIZ 18.3

1. $s = 17$ cm
2. 4.9 m
3. 1.0 m
4. 1.5×10^{-6} m

END-OF-CHAPTER QUESTIONS

1. (a) See Figure 10. (b) $s' = -19$ cm, $m = 2.9$, virtual image

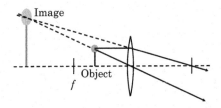

FIGURE 10   End-of-chapter questions, Chapter 18, problem 1(a).

2. (a) See Figure 11. (b) $s' = 23$ cm, $m = -1.3$, real image

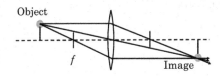

FIGURE 11   End-of-chapter questions, Chapter 18, problem 2(a).

3. $s' = 50$ cm, $m = -5.0$, real image, inverted
4. (a) 18 mm, (b) 1.7×10^{-6} m
5. (a) 4.8 ly, (b) 0.92°

ADDITIONAL PROBLEMS

1. $s' = 60$ cm, $m = -3.0$
2. $s' = -200$ mm (to the left of lens 2), magnification $= -40$
3. See Figure 12.

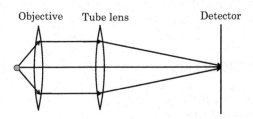

FIGURE 12   Additional problems, Chapter 18, problem 3.

4. $s' = -5.8$ mm, $m = 0.42$

CHAPTER 19

QUIZ 19.1

1. 7281 m/s
2. $c = 3.0 \times 10^8$ m/s
3. 21.5 m/s
4. $c = 3.0 \times 10^8$ m/s

QUIZ 19.2

1. $0.14\,c$
2. 3.5×10^7 s $= 1.1$ years
3. 1.4×10^8 s $= 4.6$ years
4. 4.1×10^{-6} s

QUIZ 19.3

1. 1.5×10^{-10} J
2. $K = 2.1 \times 10^{-13}$ J, $\frac{1}{2}mv^2 = 3.8 \times 10^{-14}$ J
3. 2.2×10^{-19} kg m/s
4. 9.7×10^{-19} kg m/s

END-OF-CHAPTER QUESTIONS

1. (a) 0.25 s, (b) 0.25 s, (c) 29 m/s
2. (a) 3.3×10^{-7} s, (b) 20 m $\times 10^{-6}$ s, (c) $c = 3.0 \times 10^8$ m/s
3. 49.8 years
4. (a) 8.2×10^{-14} J, (b) 1.1×10^{-13} J, (c) 2.8×10^{-14} J, (d) 2.4×10^{-22} kg m/s
5. (a) 7.3×10^8 m/s, (b) 3.0, (c) 2.9×10^8 m/s

ADDITIONAL PROBLEMS

1. (a) 60 m, (b) 1.0×10^{-7} s
2. (a) 1.0×10^{-5} s, (b) 60 m
3. 6.8×10^5 m/s, x direction
4. 2.8×10^8 m/s

CHAPTER 20

QUIZ 20.1

1. 1.1×10^5 m/s
2. -3.0 eV
3. $E = -0.85$ eV, $r = 8.5$ Å
4. 3.4 eV

QUIZ 20.2

1. 4.6×10^{14} Hz
2. 1.2×10^{-7} m $= 120$ nm
3. 126
4. ^{16}N: 7 protons, 9 neutrons, 7 electrons; ^{16}O: 8 protons, 8 neutrons, 8 electrons

QUIZ 20.3

1. 18 MeV
2. ^{60}Ni, 2.8 MeV
3. 8.3×10^{-9}
4. 13 Bq

END-OF-CHAPTER QUESTIONS

1. $E = -0.026$ eV, $r = 280$ Å
2. (a) 12.8 eV, (b) 1.6×10^{14} Hz, (c) 1.9×10^{-6} m $= 1900$ nm
3. (a) 94 protons, 145 neutrons, 94 electrons; (b) 5.2 MeV
4. 1.7 MeV
5. (a) 3.2×10^{10} Bq, (b) 3.0×10^{16}

ADDITIONAL PROBLEMS

1. (a) 4.7 eV, (b) 1.3 eV
2. 54 eV
3. 1.3 MeV
4. (a) 85 Bq, (b) 0.78 MeV

Physical Constants

DATA

Gravitational acceleration	$g = 9.80$ m/s^2
Gravitational constant	$G = 6.67 \times 10^{-11}$ N·m^2/kg^2
Mass of Earth	5.97×10^{24} kg
Radius of Earth	6.38×10^6 m
Speed of sound (in air, 20°C)	343 m/s
Gas constant	$R = 8.31$ J/(mol·K)
Mole	6.02×10^{23} molecules
Density of water	1000 kg/m^3
Coulomb's law constant ($1/4\pi\varepsilon_0$)	$K = 8.99 \times 10^9$ N·m^2/C^2
Charge of electron	-1.60×10^{-19} C
Charge of proton	1.60×10^{-19} C
Mass of electron	9.11×10^{-31} kg
Mass of proton	1.67×10^{-27} kg
Permittivity constant	$\varepsilon_0 = 8.85 \times 10^{-12}$ C^2/(N·m^2)
Permeability constant	$\mu_0 = 4\pi \times 10^{-7}$ T·m/A
Speed of light in vacuum	$c = 3.00 \times 10^8$ m/s
Planck's constant	$h = 6.63 \times 10^{-34}$ J·s

CONVERSIONS

Temperature	$T\,(\text{K}) = T\,(^\circ\text{C}) + 273$
Pressure	$1\ \text{atm} = 1.013 \times 10^5\ \text{Pa} = 101.3\ \text{kPa} = 760\ \text{mm Hg}$
Time	$1\ \text{year} = 3.16 \times 10^7\ \text{s}$
	$1\ \text{day} = 24\ \text{hours}$
	$1\ \text{hour} = 3600\ \text{s}$
	$1\ \text{minute} = 60\ \text{s}$
Energy	$1\ \text{eV} = 1.60 \times 10^{-19}\ \text{J}$
Distance	$1\ \text{Angstrom}\ (\text{Å}) = 10^{-10}\ \text{m}$
Mass	$1\ \text{atomic mass unit}\ (u) = 1.660539 \times 10^{-27}\ \text{kg}$
	$uc^2 = 931.5\ \text{MeV}$

COMMON PREFIXES

Prefix	Meaning	Example
pico (p)	10^{-12}	$1\ \text{pC} = 10^{-12}\ \text{C}$
nano (n)	10^{-9}	$100\ \text{nm} = 10^{-7}\ \text{m}$
micro (μ)	10^{-6}	$1\ \mu\text{C} = 10^{-6}\ \text{C}$
milli (m)	10^{-3}	$1\ \text{mm} = 10^{-3}\ \text{m}$
centi (c)	10^{-2}	$100\ \text{cm} = 1\ \text{m}$
kilo (k)	10^{3}	$1\ \text{g} = 10^{-3}\ \text{kg}$
mega (M)	10^{6}	$1\ \text{M}\Omega = 10^{6}\ \Omega$
giga (G)	10^{9}	$1\ \text{GW} = 10^{9}\ \text{W}$

Index